· THE OXFORD SHERLOCK HOLMES ·

General Editor
Owen Dudley Edwards

The Sign
of the Four

ARTHUR CONAN DOYLE

The Sign of the Four

Edited with an Introduction by
Christopher Roden

Oxford   New York

OXFORD UNIVERSITY PRESS

1993

Oxford University Press, Walton Street, Oxford OX2 6DP

Oxford New York Toronto
Delhi Bombay Calcutta Madras Karachi
Kuala Lumpur Singapore Hong Kong Tokyo
Nairobi Dar es Salaam Cape Town
Melbourne Auckland Madrid

and associated companies in
Berlin Ibadan

Oxford is a trade mark of Oxford University Press

British Library Cataloguing in Publication Data

Data available

Library of Congress Cataloging in Publication Data

Data available

ISBN 0-19-212316-5
ISBN 0-19-212329-7 (set)

1 3 5 7 9 10 8 6 4 2

Typeset by Pure Tech Corporation, Pondicherry, India
Printed in Great Britain by
BPCC Paperbacks Ltd
Aylesbury, Bucks

CONTENTS

ACKNOWLEDGEMENTS

THE Editor of this volume would like to thank the following persons and institutions for their encouragement
and aid in its preparation: The National Library of Scotland; Catherine Cooke, Westminster Libraries, administra-
tor of the Sherlock Holmes Collection of the Marylebone
Library; the late David G. Kirby; Christopher Redmond;
the Revd Frederick J. Turner, SJ, Archivist, Stonyhurst
College, Lancashire; Cameron Hollyer, formerly custodian
of the Arthur Conan Doyle Collection, the Metropolitan
Toronto Reference Library, Toronto, Ontario, Canada,
and the Library itself; the Manuscripts Division, Tracy W.
McGregor Library, Special Collections Department,
University of Virginia Library, Charlottesville, Virginia,
USA; the William Hill Organization; Alfred Dunhill Ltd;
Douglas S. Warren; the Stormy Petrels of British Columbia,
Canada.

And, above all, Barbara Roden.

The General Editor's thanks for obligations incurred
by the entire series are asserted in the acknowledgements to
A Study in Scarlet.

GENERAL EDITOR'S PREFACE
TO THE SERIES

ARTHUR CONAN DOYLE told his *Strand* editor, Herbert Greenhough Smith (1855–1935), that 'A story always comes to me as an organic thing and I never can recast it without the Life going out of it.'[1]

On the whole, this certainly seems to describe Conan Doyle's method with the Sherlock Holmes stories, long and short. Such manuscript evidence as survives (approximately half the stories) generally bears this out: there is remarkably little revision. Sketches or scenarios are another matter. Conan Doyle was no more bound by these at the end of his literary life than at the beginning, whence scraps of paper survive to tell us of 221B Upper Baker Street where lived Ormond Sacker and J. Sherrinford Holmes. But very little such evidence is currently available for analysis.

Conan Doyle's relationship with his most famous creation was far from the silly label 'The Man Who Hated Sherlock Holmes': equally, there was no indulgence in it. Though the somewhat too liberal Puritan Micah Clarke was perhaps dearer to him than Holmes, Micah proved unable to sustain a sequel to the eponymous novel of 1889. By contrast, 'Sherlock' (as his creator irreverently alluded to him when not creating him) proved his capacity for renewal 59 times (which Conan Doyle called 'a striking example of the patience and loyalty of the British public'). He dropped Holmes in 1893, apparently into the Reichenbach Falls, as a matter of literary integrity: he did not intend to be written off as 'the Holmes man'. But public clamour turned Holmes into an economic asset that could not be ignored. Even so, Conan Doyle could not have continued to write about

[1] Undated letter, quoted by Cameron Hollyer, 'Author to Editor', *ACD— The Journal of the Arthur Conan Doyle Society*, 3 (1992), 19–20. Conan Doyle's remark was probably *à propos* 'The Red Circle' (*His Last Bow*).

Holmes without taking some pleasure in the activity, or indeed without becoming quietly proud of him.

Such Sherlock Holmes manuscripts as survive are frequently in private keeping, and very few have remained in Britain. In this series we have made the most of two recent facsimiles, of 'The Dying Detective' and 'The Lion's Mane'. In general, manuscript evidence shows Conan Doyle consistently underpunctuating, and to show the implications of this 'The Dying Detective' (*His Last Bow*) has been printed from the manuscript. 'The Lion's Mane', however, offers the one case known to us of drastic alterations in the surviving manuscript, from which it is clear from deletions that the story was entirely altered, and Holmes's role transformed, in the process of its creation.

Given Conan Doyle's general lack of close supervision of the Holmes texts, it is not always easy to determine his final wishes. In one case, it is clear that 'His Last Bow', as a deliberate contribution to war propaganda, underwent a ruthless revision at proof stage—although (as we note for the first time) this was carried out on the magazine text and lost when published in book form. But nothing comparable exists elsewhere.

In general, American texts of the stories are closer to the magazine texts than British book texts. Textual discrepancies, in many instances, may simply result from the conflicts of sub-editors. Undoubtedly, Conan Doyle did some rereading, especially when returning to Holmes after an absence; but on the whole he showed little interest in the constitution of his texts. In his correspondence with editors he seldom alluded to proofs, discouraged ideas for revision, and raised few—if any—objections to editorial changes. For instance, we know that the *Strand*'s preference for 'Halloa' was not Conan Doyle's original usage, and in this case we have restored the original orthography. On the other hand, we also know that the *Strand* texts consistently eliminated anything (mostly expletives) of an apparently blasphemous character, but in the absence of manuscript confirmation we have normally been unable to restore what were probably stronger original versions. (In any case, it is perfectly possible that Conan Doyle, the consummate professional, may have come to exercise self-censorship in the certain knowledge that editorial changes would be imposed.)

Throughout the series we have corrected any obvious errors, though these are comparatively few: the instances are at all times noted. (For a medical man, Conan Doyle's handwriting was commendably legible, though his 'o' could look like an 'a'.) Regarding the order of individual stories, internal evidence makes it clear that 'A Case of Identity' (*Adventures*) was written before 'The Red-Headed League' and was intended to be so printed; but the 'League' was the stronger story and the *Strand*, in its own infancy, may have wanted the series of Holmes stories established as quickly as possible (at this point the future of both the Holmes series and the magazine was uncertain). Surviving letters show that the composition of 'The Solitary Cyclist' (*Return*) preceded that of 'The Dancing Men' (with the exception of the former's first paragraph, which was rewritten later); consequently, the order of these stories has been reversed. Similarly, the stories in *His Last Bow* and *The Case-Book of Sherlock Holmes* have been rearranged in their original order of publication, which—as far as is known—reflects the order of composition. The intention has been to allow readers to follow the fictional evolution of Sherlock Holmes over the forty years of his existence.

The one exception to this principle will be found in *His Last Bow*, where the final and eponymous story was actually written and published after *The Valley of Fear*, which takes its place in the Holmes canon directly after the magazine publication of the other stories in *His Last Bow*; but the removal of the title story to the beginning of the *Case-Book* would have been too radically pedantic and would have made *His Last Bow* ludicrously short. Readers will note that we have already reduced the extent of *His Last Bow* by returning 'The Cardboard Box' to its original location in the *Memoirs of Sherlock Holmes* (after 'Silver Blaze' and before 'The Yellow Face'). The removal of 'The Cardboard Box' from the original sequence led to the inclusion of its introductory passage in 'The Resident Patient': this, too, has been returned to its original position and the proper opening of 'The Resident Patient' restored. Generally, texts have been derived from first book publication collated with magazine texts and, where possible, manuscripts; in the case of 'The Cardboard Box' and 'The Resident Patient', however, we have employed the *Strand* texts, partly because of the restoration of the latter's opening, partly to give readers a flavour of the magazine in which the Holmes stories made their first, vital conquests.

In all textual decisions the overriding desire has been to meet the author's wishes, so far as these can be legitimately ascertained from documentary evidence or application of the rule of reason.

One final plea. If you come to these stories for the first time, proceed now to the texts themselves, putting the introductions and explanatory notes temporarily aside. Our introductions are not meant to introduce: Dr Watson will perform that duty, and no one could do it better. Then, when you have mastered the stories, and they have mastered you, come back to us.

OWEN DUDLEY EDWARDS

University of Edinburgh

INTRODUCTION

THE stories of Sherlock Holmes—which were, and remain, the most popular works ever written by Sir Arthur Conan Doyle (1859–1930)—appeared over forty years between 1887 and 1927. The period saw their author rise from a relatively impecunious medical general practice to a place among the highest-paid writers of his day. He was a young doctor struggling to make ends meet in private practice in Southsea when *A Study in Scarlet* was published in 1887. The story had been submitted to Ward, Lock & Co. in September 1886 and the publishers accepted it on certain conditions, one being that it would be held over until the following year in view of the quantity of cheap literature that was on the market at the time. Conan Doyle, no doubt anxious to see his work in print, agreed to the publishers' terms and signed away all future British rights to the story for the meagre sum of £25. He was, as he had been warned, to wait a full year before the story was published, and its eventual appearance attracted little immediate public attention: the character of Sherlock Holmes was to create no great early impression.

Conan Doyle persevered, pursuing his writing career alongside the call of his full-time medical duties. His *The Mystery of Cloomber* was published in early 1888, but even at that early stage he was dissatisfied with its sensational nature: his inclination was to emulate the works of his literary mentors Sir Walter Scott and Thomas Babington Macaulay. In his autobiography, *Memories and Adventures* (1924), he wrote: 'I now determined to test my powers to the full, and I chose a historical novel for this end, because it seemed to me the one way of combining a certain amount of literary dignity with those scenes of action and adventure which were natural to my young and ardent mind.' The result of his further endeavours was *Micah Clarke*, written in 1887. The difficulties that he had experienced in trying to place the manuscript of *A Study in Scarlet* were to be repeated when *Micah Clarke* started its round of the publishing-houses. James Payn, the editor of *Cornhill Magazine*, rejected it, as did Richard Bentley & Son (who had published some of his work in their magazine, *Temple Bar*) and William Blackwood and Sons (who had refused him in theirs). But in September 1888 he sent the manuscript to Longman's, where it came to the attention of Andrew Lang (1844–1912). Lang liked *Micah Clarke*, advised its acceptance, and Conan Doyle signed a contract on 29 October 1888. Once the book was published, it received, in his own words, 'extraordinarily good reviews', and sales were impressive. Lang, writing in the *Quarterly Review* for July 1904, commented: '*Micah Clarke* is a long novel of five hundred and seventy pages; but nobody, when he has finished it, remembers that it is long—which is praise enough for any romance.' Its publication in February 1889 followed another happy event in Conan Doyle's life. On 28 January he and his wife, Louisa ('Touie'), whom he had married in August 1885, became parents for the first time with the birth of their daughter, Mary Louise. Conan Doyle's literary output took on a new significance: his income as a doctor, which never exceeded £300, was supplemented by Louisa's small private income, but there was now a family to support. During the Easter of 1889 he began extensive readings which were to form the basis of his new historical novel, and his diary shows that he began work on *The White Company* on 19 August 1889. The novel threw him into late fourteenth-century England, France, and Spain in place of *Micah Clarke*'s Monmouth revolt in the west country of 1685; but for all the heavy research it entailed it would no doubt have been finished much earlier than June 1890 had Conan Doyle not been invited to dine in London at the end of August 1889. It was the single event that determined, more than any other, the shape of what he was to write for the next 38 years.

Joseph Marshall Stoddart (1845–1921) had rejoined J. B. Lippincott and Co., the leading publishers in his native Philadelphia, and as managing editor of *Lippincott's Magazine* was to launch an English edition, with an English editor, English contributors, and, above all, English sales. His firm linked up with Ward, Lock for British publication, and from their staff he acquired their chief editor or principal reader, George Thomas Bettany (1850–91), to be English editor with his new junior editor, John Coulson Kernahan (1858–1943), as Assistant English editor. Bettany, by profession a scientist (he had relinquished his Lectureship in Botany at Guy's Hospital in 1886 to join Ward, Lock), was ready enough to edit libraries of 'Famous Books' in literary and historical as well as scientific fields, but he seems to have been at more of a loss with contemporary writing. Though *A Study in Scarlet* had been accepted by Ward, Lock on his advice he had in fact acted on his wife's. The first two volumes of *Lippincott's* 'Special English Edition' duly included Jeanie Gwynne Bettany's 95-page *A Laggard in Love* (a prolific novelist later, so far she had produced only one three-volume novel and one sketch of schoolboys' home life), the anonymous serial *'A Dead Man's Diary'* (actually by Coulson Kernahan), and A. Conan Doyle's *The Sign of the Four*— seemingly the limit of Ward, Lock's offering of English novelists to *Lippincott's* for 1890. Oscar Wilde was provided by Stoddart, who had sponsored his Philadelphia lectures in 1882, while the presence of William Clark Russell (1844–1911), whose *A Marriage at Sea* supplied the October novel, may well have been the result of a suggestion over the dinner-table by his enthusiastic admirer Conan Doyle. For the rest, the English edition's monthly novels of 1890 were by Americans well known in England such as Julian Hawthorne (1846–1934), who in addition to his *Millicent and Rosalind* contributed a serial four-part discussion of a story draft by his father Nathaniel (1804–64) with supporting notebook material which must have greatly interested Conan Doyle, a future beneficiary of the literary influence of Nathaniel Hawthorne; and the late US Consul to Glasgow, Francis Bret Harte (1836–1902), the high noon of whose influence on Conan Doyle had passed with *A Study in Scarlet*; and other Americans whose English reputation was less clearly defined, such as Captain Charles King (1844–1933), Christian Reid (1846–1920), Kate Pearson Woods (1853–1932), and Mary Etta Stickney (1853–?). The birthdates are suggestive: Conan Doyle was the youngest contributor. With so little to show for Ward, Lock's own efforts, it is understandable that Bettany (and his wife) should put forward their one conspicuous literary discovery, and equally understandable that Joseph M. Stoddart would want to take a look at him before deciding anything. And he had also to satisfy himself that there was a good short novel in his exotic friend Wilde.

This view of the invitation will be new to Conan Doyle students. Hitherto James Payn (1830–98), editor of the *Cornhill*, who had published a couple of Conan Doyle short stories, has been given credit for the introduction; but it seems much more likely that the sequel to *A Study in Scarlet* was induced on the recommendation of the editor who had accepted it, rather than the one who had rejected it. Payn may have given a supporting testimonial; but Bettany needed to produce a name. It would, in any case, have been too ironic had Payn helped draw up the guest-list for a dinner including Wilde, who some eight months earlier, in 'The Decay of Lying', had told *Nineteenth Century* readers: 'Mr James Payn is an adept in the art of concealing what is not worth finding. He hunts down the obvious with the enthusiasm of a short-sighted detective. As one turns the pages, the suspense of the author becomes unbearable.' And detectives, of the keener-sighted variety, were very much on the organizers' minds. The detective story had been on the verge of takeoff for the last couple of years. All the works of Émile Gaboriau (1832–73) were becoming available, even in rival translations. *The Mystery of a Hansom Cab*, by Fergus Hume (1859–1932), published in 1886, was an all-time bestseller. Julian Hawthorne, slated for January 1890 to lead the *Lippincott* 'English edition' novels, was himself producing such Cassell publications as *A Tragic Mystery* (1888), *Section 558: or, the Fatal Letter* (1888), and *Another's Crime* (1889), all supposedly 'From the Diary of Inspector Byrnes of the Detective Bureau'. The inference is clear enough: the Bettanys would seem to have conveyed to Stoddart not only the author but, if possible, the character, to commission. They had brought Holmes to print, and they wanted more Holmes.[1]

The meeting of Stoddart and Conan Doyle took place in the luxurious surroundings of the Langham Hotel, as Conan Doyle recalled with some pleasure in *Memories and Adventures*:

Stoddart, the American, proved to be an excellent fellow, and had two others to dinner. They were [Thomas Patrick] Gill [1858–1931], a very entertaining Irish M.P. [associate editor of the *North American Review* 1883–5, probably present as Stoddart's private assessor], and Oscar Wilde [1854–1900], who was already famous as the champion of aestheticism. It was indeed a golden evening for me. Wilde to my surprise had read *Micah Clarke*, and was enthusiastic about it, so that I did not feel a complete outsider. His conversation left an indelible impression upon my mind. He towered above us all, and yet had the art of seeming to be interested in all that we could say. He had delicacy of feeling and tact, for the monologue man, however clever, can never really be a gentleman at heart. He took as well as gave, but what he gave was unique. He had a curious precision of statement, a delicate flavour of humour, and a trick of small gestures to illustrate his meaning, which were peculiar to himself. The effect cannot be reproduced . . .

The result of the evening was that both Wilde and I promised to write books for *Lippincott's Magazine*—Wilde's contribution was *The Picture of Dorian Gray*, a book which is surely upon a high moral plane, while I wrote *The Sign of Four* in which Holmes made his second appearance. . . .

---

[1] The story that Jeanie Gwynne Bettany, who after her husband's death became Mrs Coulson Kernahan, soured on Holmes is baseless; she has been confused with Frederick George Bettany (1868–1942), Literary Editor of the *Sunday Times* 1901–17. Her second husband's account of her midwifery to *A Study in Scarlet* was published in her lifetime (d. 1941) and obviously with her permission, drawing on her recollection (Coulson Kernahan, 'Personal Memories of Sherlock Holmes', *London Quarterly and Holborn Review*, Oct. 1934, 449–50). Kernahan's impassioned post-war denunciations of Conan Doyle for his Spiritualist beliefs made any allusion to the Kernahans in *Memories and Adventures* an improbability: Conan Doyle may never have known about Jeanie Bettany Kernahan's role.

When his little book came out I wrote to say what I thought of it. His letter ... comments on my own work in too generous terms.

Conan Doyle's contract, dated 30 August 1889, may have been drawn up on the same day as the meal itself. It provided that *Lippincott's* would pay £100 for a story of not less than 40,000 words. The magazine was to purchase full American rights (along with three months' rights in England), and the manuscript was to be delivered before January 1890. This had involved at least one hard word. Lippincott's were the American publishers, of course, but Ward, Lock wanted to pick up the British publication of novels previously appearing in *Lippincott's*. Oscar Wilde, for instance, gave them *The Picture of Dorian Gray*, and bitterly regretted it, having been told that their 10 per cent was an invariable royalty only to discover it was not. Conan Doyle, having failed to get anything more than £25 for the British copyright of *A Study in Scarlet*, absolutely refused to surrender his rights beyond the three months, arranged for serialization after that time in provincial newspapers, and had his book published in October 1890 by Spencer Blackett: two years later George Newnes (by then most successfully publishing Holmes in the *Strand* and in the book collection of the *Adventures*) took over the title. The cause of Conan Doyle's hostility to Ward, Lock may have been the same as Wilde's—the hard-bitten James Bowden, whose financial zeal encouraged few geese to lay more golden eggs than one in the Ward, Lock nest. Book people found their gains dissipated by book-keepers, and Bowden got his partnership while Bettany died.

But good relations matured happily between Conan Doyle and Joseph Marshall Stoddart, and the new Lippincott author was writing on 3 September 1889:

As far as I can see my way at present my story will either be called 'The Sign of the Six' or 'The Problem of the Sholtos'. You said you wanted a spicy title. I shall give Sherlock Holmes of *A Study in Scarlet* something else to unravel. I notice that everyone who has read the book wants to know more of that young man. Of course the new story will be entirely independent of *A Study in Scarlet* but as Sherlock Holmes & Dr Watson are introduced in each, I think that the sale of one might influence the other. I wish therefore that your firm would reprint *A Study in Scarlet* in America and give me some dollars for it.

In *Memories and Adventures* he recalled American piracies but dated some before Stoddart's mission, which he assumed was prompted by *A Study in Scarlet*'s illicit successes. But Lippincott's publication in response to Conan Doyle's Plea is the first American edition of *A Study in Scarlet* so far as is known, and Stoddart's knowledge of the book's success was gained in London. The letter would seem a contract of content, summing up the discussion, recalling on reflection how well casual readers had reacted to Holmes, and seeing his way to some compensation for his losses by Ward, Lock—and Bowden. There is nothing to indicate whether or not Conan Doyle had thought of a sequel to *A Study in Scarlet* before the Langham meeting. At the end of 1888 he had worked on a three-act drama entitled 'Angels of Darkness', which was based on the Mormon episode in that story's second Part. The script, which has never been produced or published, included Dr Watson but excluded Sherlock Holmes. Whatever Conan Doyle's intention, *The Sign of the Four* presented him with the opportunity to reconstruct Holmes and reintroduce the detective to the reading public.

The Sign of the Four was completed with considerable speed. Conan Doyle's diary for 30 September 1889 records: ' "The Sign of the Four" finished and dispatched.' The manuscript, some 160 pages long, was forwarded to *Lippincott's*, although it appears that Conan Doyle had still not settled the title in his own mind. On 1 October he wrote to Stoddart:

You promised to collaborate with me in this book, so I want you to name it, which will surely make you a bona fide collaborator. 'The Sign of the Four' strikes me as likely to be popular, but a trifle catchpenny. 'The Problem of the Sholtos' is more choice, though less dramatic. On you be the burden of choosing. I wish also that your reader would look over the proofs. It would save sending twice across the Atlantic.

I have finished & sent it to your London agents. I think it is pretty fair, though I am not usually satisfied with my own things. It has the advantage over the *Study in Scarlet* not only as being much more intricate, but also as forming a connected narrative without any harking back as in the second part of the *Study*. Holmes, I am glad to say, is in capital form all through. You will see that the *Study* is alluded to in it, and I believe that you have a demand for it if you would get it out at the same time.[2]

Stoddart adopted the simple course and used both titles for the story's first publication, in magazine and almost simultaneous book appearances in February 1890: *The Sign of the Four; or, The Problem of the Sholtos,* with designation of A. Conan Doyle as author of *Micah Clarke* and of *A Study in Scarlet* respectively gratifying his intellectual and his economic aspirations. English publication opted for *The Sign of Four* from the first, and Conan Doyle—if he thought about it at all—may have come to prefer this form. But throughout the story the reference is to 'the sign of the four' and that version is, therefore, employed here.

The speed of *The Sign of the Four*'s composition reflects ready sources of its inspiration. First, there was India. Before coming to Portsmouth, Conan Doyle had produced ill-informed, far-flung juvenilia, gravely sub-titled 'true colonial story' and the like, and happily set in locations from Arizona to Australia: he blushed at their memory later. But his writings on India from the first reflected an informed guidance, dating as they do from the years after settlement at Southsea and acquisition of Major-General Alfred Wilks Drayson (1827–1901) as patient, mentor, and sponsor at the Portsmouth Literary and Scientific Society. Drayson commanded the 21st Brigade Artillery in India from 1876 to 1878 (and Artillerymen were thus automatic objects for detection by Sherlock and Mycroft Holmes in 'The Greek Interpreter', *Memoirs*); and

[2] Quotations from Conan Doyle/Stoddart letters from *Uncollected Sherlock Holmes*, 494–50.

. . . it having been considered desirable, that alterations should be made, both in the details, and rearmament, of various forts in the Bengal Presidency, I was appointed president of the committee ordered to assemble for the purpose. Allahabad was the first fort I inspected and reported on; after this I visited Agra, Gwalior, Delhi, and Fort William, Calcutta. The suggestions I made, were all carried out, and I have reason to believe that I saved the Indian Government many lacs of rupees.[3]

Conan Doyle had only to look in on his patient to get exact detail on Agra.

Conan Doyle himself was conferred MD at the Edinburgh University graduation on 1 August 1885 by the new Principal, Sir William Muir (1819–1905), and if they talked then or later (not at all impossible: Muir was the most accessible Principal that students had known in living memory), he could have heard of Muir's own work on the defences of Agra during the Mutiny—accounts of Muir at his installation as Principal should have mentioned the fact. Drayson would have asked Conan Doyle who would confer him, and would have known Muir's name because of his own study of the Agra defences: and from there the old soldier's 'Well, if you get half a chance, ask the fella . . .' seems inevitable. Like Tennyson's old Ulysses, the modern Major-General was a part of all that he had met and evidently intended his circle of friends to be part of it as well.

The Drayson connection with the Agra defences is of more consequence than Conan Doyle's dependence on *The Moonstone* (1868) by William Wilkie Collins (1824–89) in writing *The Sign of the Four*. Both novels are concerned with the theft of human treasure (the sacred yellow diamond in *The Moonstone*, a hoard of precious stones and pearls in the *Sign*) bringing revenge from India to Britain for evil actions of the soldiers of the British Empire: there are comparable glimpses of the initial murder, Collins bringing us in just after it, Conan Doyle letting us witness it through the

---

[3] Drayson, *Experiences of a Woolwich Professor during Fifteen Years at the Royal Military Academy* (1886), 305–6.

murderer's eyes. But there the matter ends. Three Indians recover the Moonstone; three Indians want to recover the treasure in *The Sign of the Four*, but the only one of the four Signatories to do so is white. *The Moonstone* had been by far a greater influence on Conan Doyle's *The Mystery of Cloomber* (1888), with its three Indian holy men whose revenge journey also ends in the death of the defiler of their religion (in this case General John Berthier Heatherstone for the murder of an ancient saint). Both Drayson and Conan Doyle were immersed in Theosophy in 1883 (which is when *The Mystery of Cloomber* seems to have been written), although both of them then became suspicious of it. The result is that in the later work the idea of a spiritual pursuit from an avenging Indian trio remains, but instead of subjecting the credulity of the reader to astral bodies or deriving too obviously from Collins, *The Sign of the Four*, by concentrating on Jonathan Small's comradeship for his non-white co-signatories, makes this loyalty one of his most attractive qualities. Small's story falls into its natural place in *The Sign of the Four* with no necessity to borrow Collins's multi-narrator technique as in *The Mystery of Cloomber*, or with a narrative of earlier events of uncertain provenance as in *A Study in Scarlet*.

Sergeant Cuff, of *The Moonstone*, has been given some credit as an ancestor of Sherlock Holmes; but Cuff, while a detective working alone, is more representative of the official police force as encountered in the Holmes stories in the persons of Lestrade, Gregson, and Athelney Jones. Cuff's contempt for Superintendent Seegrave might seem Holmesian (and is startlingly open given the difference in their ranks, however much London might lord it over the provinces), but its strongest effect is the mutual jealousy of Gregson and Lestrade in *A Study in Scarlet*. Athelney Jones in *The Sign of the Four* may owe a touch or two to Seegrave, but there is nothing between Cuff and Seegrave anywhere as memorable as:

'. . . Ha! I have a theory. These flashes come upon me at times . . . What do you think of this, Holmes? Sholto was, on his own confession, with his brother last night. The brother died in a fit, on which Sholto walked off with the treasure! How's that?'

'On which the dead man very considerately got up and locked the door on the inside.'

'Hum! There's a flaw there . . .'

Cuff's modesty may be compared to that of Holmes, who on many occasions was prepared to allow the police force to take the credit for his own successes. From Cuff, too, may have come Holmes's penchant for asking a seemingly trivial question on which great things ultimately prove to turn. Conan Doyle may also have been influenced by two other aspects of *The Moonstone*. The opium addiction of Ezra Jennings and the drugging of Franklin Blake (plus his own practical knowledge of drugs for medical uses) may have decided him in favour of making Holmes resort to cocaine (and, says Watson, morphine) as a stimulant for relieving *ennui*. And the character of Gooseberry seems one origin for the Baker Street Irregulars (also for Cartwright, in *The Hound of the Baskervilles*). The Irregulars, Holmes's unofficial detective police force, first appeared in *A Study in Scarlet* but were used much more obtrusively in *The Sign of the Four*. Collins's prominently-eyed Gooseberry was, like the Baker Street Irregulars, ubiquitous:

'. . . Did you notice my boy—on the box there?'

'I noticed his eyes.'

Mr Bruff laughed. 'They call the poor little wretch "Gooseberry" at the office', he said. 'I employ him to go on errands—and I only wish my clerks who have nick-named him were as thoroughly to be depended on as he is. Gooseberry is one of the sharpest boys in London.' (*The Moonstone*, Second Period, Fifth Narrative, Chapter 1)

Ex-sergeant Cuff later remarks that 'One of these days . . . that boy will do great things in my late profession. He is the brightest and cleverest little chap I have met with for many a long year past.' We last see him, dancing with delight, while Cuff washes a corpse's face to identify the villain, as Holmes would do to 'The Man With the Twisted Lip' (*Adventures*).

Edgar Allan Poe (1809–49) was described by Conan Doyle in *Through the Magic Door* as 'to my mind, the supreme original short story writer of all time'. In *The Sign of the Four* the presence of Poe can perhaps be detected in the grotesque figure of the tiny Andaman Islander, Tonga—'a little black man—the smallest I have ever seen—with a great, misshapen head and a shock of tangled, dishevelled hair'. Tonga is a human endowed with bestial characteristics: 'His small eye glowed and burned with a sombre light, and his thick lips were writhed back from his teeth, which grinned and chattered at us with half animal fury.' He is the reverse, in fact, of the orang-utan in Poe's 'The Murders in the Rue Morgue' (1841), who is a beast presented initially with human attributes ('Razor in hand, and fully lathered, it was sitting before a looking-glass, attempting the operation of shaving, in which it had no doubt previously watched its master through the key-hole of the closet').

The most overlooked influence on *The Sign of the Four* is Conan Doyle's fellow Scot Robert Louis Stevenson (1850–94). Conan Doyle wrote of Stevenson in *Through the Magic Door*:

There is much more that might be said as to Stevenson's original methods in fiction. As a minor point it might be remarked that he is the inventor of what may be called the mutilated villain. It is true that Mr Wilkie Collins has described one gentleman who had not only been deprived of all his limbs, but was further afflicted by the insupportable name of Miserrimus Dexter. Stevenson, however, has used the effect so often, and with such telling results, that he may be said to have made it his own. To say nothing of Hyde, who was the very impersonation of deformity, there is the horrid blind Pew, Black Dog with two fingers missing, Long John with his one leg.

Conan Doyle would seem to have left Black Dog where he found him, and for a child of Pew in sheer terrorizing effect we must go beyond *The Sign of the Four*: the giggling Wilson Kemp in 'The Greek Interpreter' (*Memoirs*) may be the nearest approximation. But *Treasure Island* was in Conan Doyle's mind in writing *The Sign of the Four*. His original working title, 'The Sign of the Six', recalls the six in Stevenson's plot: the six who accompanied Flint when he went ashore to bury the treasure—and who never returned. The skeleton of one is a sign in itself: 'Flint's Pointer'. Conan Doyle's six whom the Agra Treasure enslaves and destroys would have included Sholto and Morstan—or some variant of them—and 'the sign of the four' as placed on Major Sholto's corpse has echoes of the 'Black Spot' delivered by Pew to Billy Bones, and later by George Merry to Long John Silver, a formal denunciation of a former comrade. Billy Bones's terror of a one-legged man prefigures Sholto's 'most marked aversion to men with wooden legs. On one occasion he actually fired his revolver at a wooden-legged man, who proved to be a harmless tradesman canvassing for orders. We had to pay a large sum to hush the matter up.'

But the transition from superstitious eighteenth-century Cornish boy to sophisticated nineteenth-century London man is marked by the contrast of Jim Hawkins's hideous nightmares and Thaddeus Sholto's cool 'My brother and I used to think this a mere whim of my father's; but events have since led us to change our opinion', eccentricities in any case offering less basis for omens when an accustomed paternal habit. Small himself is a very different character from Silver, being essentially an unfortunate person rather than an evil one; he displays human sympathy, understanding, and above all loyalty, qualities which Stevenson's charismatic, cold-hearted pirate so conspicuously lacks. But both Small and Silver use their disability to murderous effect: Silver breaks Tom's back with his thrown crutch while Small with his wooden leg 'knocks the whole front' of his hated guard's skull in: 'You can see the split in the wood now where I hit him.' Yet while Hawkins's horrified sight of the murder of Tom is our most atrocious glimpse of Silver, Conan Doyle, who wrote much closer to comedy than Stevenson, actually reprises the wooden-leg killing in a Maupassant-like mix for the final exeunt of Jonathan Small and Athelney Jones:

'Good night, gentlemen both', said Jonathan Small.

'You first, Small', remarked the wary Jones as they left the room. 'I'll take particular care that you don't club me with your wooden leg, whatever you may have done to the gentleman at the Andaman Isles.'

One last echo of Stevenson may be present in Small's irrecoverable treasure-dispersal, recalling Prince Florizel's last rites over the Rajah's Diamond in the *New Arabian Nights*: 'The Prince made a sudden movement with his hand, and the jewel, describing an arc of light, dived with a splash into the flowing river.' But while the Agra treasure, like the Diamond, has driven men to crime and treachery, Conan Doyle cuts the moral even closer to the bone than Stevenson. Jonathan Small's cry that his three Indian comrades 'would have had me do just what I have done, and throw the treasure into the Thames rather than let it go to kith or kin of Sholto or Morstan' may seem unfair to Morstan, who (so far as anyone including Small knows) acted with honour throughout. But by Small's action Watson and Mary Morstan find happiness which would have been denied to them had the treasure survived: involuntarily, Small has made a just settlement of accounts with the man who stood by him, as well as with the one who betrayed him. What is also nearer the bone than Stevenson is the justice of Small's bitter jeer: ' "You'll find the treasure where the key is, and where little Tonga is." '

Oscar Wilde must also be included among influences on *The Sign of the Four*. The description of Thaddeus Sholto is Wilde, only in one sentence:

Nature had given him a pendulous lip, and a too visible line of yellow and irregular teeth, which he strove feebly to conceal by constantly passing his hand over the lower part of his face.

And his conversation recalls Wilde only obliquely. Wilde was too kind-hearted to make a personal remark of the gross insensitivity with which Thaddeus announces Captain Morstan's death, but Mary Morstan's and Watson's reactions resembled some responses to Wilde's more outrageous conversational forays. As a talker, Thaddeus is no Wilde, but his art-chatter touches the fringes of Wilde's world; Wilde's influence made Holmes here and later more flamboyant, more epigrammatic, and, occasionally, even affected Watson whose comment on the newspaper report of Athelney Jones's bag of arrests—' "I think that we have had a close shave ourselves of being arrested for the crime" '—is gratifyingly Wildean. Conan Doyle profited from study of Wilde's dialogues such as 'The Decay of Lying' and 'The Critic as Artist' whose views are at various times adopted or refuted by Holmes, notably in the opening of 'A Case of Identity' (*Adventures*). Wilde's hand beckons Sherlock Holmes into the 'nineties.

And *The Sign of the Four* is a most appropriate product of the same paternity ward as *The Picture of Dorian Gray*. Neither Oscar Wilde nor Conan Doyle were drug-users, but Dorian Gray plays with drugs and debauches others into becoming addicts, while the Holmes of *The Sign of the Four* is, immediately and sensationally, very different from the character presented in *A Study in Scarlet*:

Sherlock Holmes took his bottle from the corner of the mantelpiece, and his hypodermic syringe from its neat morocco case. With his long, white, nervous fingers he adjusted the delicate needle and rolled back his left shirt-cuff. For some little time his eyes rested thoughtfully upon the sinewy forearm and wrist, all dotted and scarred with innumerable puncture-marks. Finally, he thrust the sharp point home, pressed down the tiny piston, and sank back into the velvet-lined armchair with a long sigh of satisfaction.

This graphic description of Holmes's dependence on cocaine (in 'seven-per-cent solution') has sometimes been omitted from editions of the novel intended for juvenile readers. Conan Doyle's Victorian audience, for whom cocaine and opium were widely available in comestibles and medicinal preparations, was certainly less sensitive than modern readers to the implications of Holmes's habit, which served as a stimulus when he was denied the 'mental exaltation' provided by his professional activities ('my own proper atmosphere'). ' "Give me problems, give me work, give me the most abstruse cryptogram, or the most intricate analysis," ' he tells Watson. ' "I can dispense then with artificial stimulants." ' This dramatic presentation of Holmes in the opening paragraphs of the novel is part of Conan Doyle's attempts to give greater depth and solidity to the character of his consulting detective. Holmes's first appearance had been in an ephemeral publication (*Beeton's Christmas Annual* for 1887) whence it had been lucky to progress into separate publication as a railway-bookstall shilling-shocker. With *The Sign of the Four* Conan Doyle wished to make his creation known to a wider and more reputable readership, and a more sophisticated presentation of Holmes than in *A Study in Scarlet* was essential to his purpose. In the earlier book he had deliberately placed limitations on Holmes's character and accomplishments. Watson's initial assessments of his new acquaintance's limits are categoric:

1. Knowledge of Literature.—Nil.
2. Knowledge of Philosophy.—Nil.
3. Knowledge of Astronomy.—Nil.
4. Knowledge of Politics.—Feeble

But in *The Sign of the Four* all this changes, and Holmes uses Euclid as a lever to deride *A Study in Scarlet* itself, cites Goethe to digest Athelney Jones, and educates Watson on Winwood Reade. The most blatant reversal of his ignorance arises from Conan Doyle's having lectured to the Portsmouth Literary and Scientific Society (19 January 1886) on 'Thomas Carlyle and his Works', a few months after which he made Watson say of Holmes in *A Study in Scarlet*:

Upon my quoting Thomas Carlyle, he inquired in the naïvest way who he might be and what he had done.

But in *The Sign of the Four*:

' . . . How small we feel, with our petty ambitions and strivings, in the presence of the great elemental forces of Nature! Are you well up in your Jean Paul [Richter]?'

'Fairly so. I worked back to him through Carlyle.'
'That was like following the brook to the parent lake . . .'

The satirical element in the portrait of Holmes in *A Study in
Scarlet*, acting out the philistinism of specialists, is dropped at
this point, in favour of a vindication of the specialist's
necessity for cultural self-nourishment. But it is accomplished with *panache*, the ignoramus on Carlyle becoming the
austere critic of his dependence on the German Romantic
achievement. New readers are intended to be impressed,
old ones shocked. In Wildean style, the inconsistency is
flourished and never explained.

Holmes emerges from the new novel as an altogether
more rounded and plausible character; but also—import-
antly—with the potential for further development. *A Study
in Scarlet* had been Conan Doyle's first full experiment in
detective fiction; in *The Sign of the Four* he exhibited a
growing confidence in his creations and in the technical
aspects of his new craft, which he was still pursuing in
conjunction with a full-time medical career. It was not only
Holmes who was given greater stature and resonance: the
figure of Dr Watson, too, benefits from Conan Doyle's
increasing creative assurance. The relationship between
these two central characters becomes an essential component
of the narrative and is based on a growing bond of friendship that was to become a key ingredient of the ensuing
cycle's success. Watson also appears at the outset of *The Sign
of the Four*, as the concerned comrade with expert knowledge
and medical responsibility to reinforce his anxieties:

'But consider!' I said earnestly. 'Count the cost! Your brain may,
as you say, be roused and excited, but it is a pathological and
morbid process which involves increased tissue-change and may at
least leave a permanent weakness. You know, too, what a black
reaction comes upon you. Surely the game is hardly worth the
candle. Why should you, for a mere passing pleasure, risk the loss
of those great powers with which you have been endowed?
Remember that I speak not only as one comrade to another but
as a medical man to one for whose constitution he is to some
extent answerable.'

The deepening of the relationship between Holmes and Watson was a striking development and opened up a wide range of narrative possibilities. In *A Study in Scarlet* the two were only lightly conjoined; in *The Sign of the Four* the relationship is more fully developed and Watson is also allowed to play an active role in his own right. Holmes was to assume the dominant position throughout the succeeding cycle of stories; but Watson was never completely over-shadowed. There was a balance in the set of one character against the other so that, for all Holmes's brilliance, the character of Watson was never diminished or ridiculed. On several occasions he even became the agent of self-rebuke in the great detective, as at the conclusion of 'The Yellow Face'. The balance of the relationship begins to be apparent in *The Sign of the Four*; for example when Holmes, after making deductions about Watson's brother, is brought to terms with the human reality that his companion's feelings have been hurt and must beg him to ' "accept my apologies. Viewing the matter as an abstract problem, I had forgotten how personal and painful a thing it might be to you." ' This has a certain condescending formality about it; but it was a marker for the future, reaching its apotheosis, perhaps, in 'The Three Garridebs' (1925), when Holmes displays un-guessed-of emotion in terror for Watson's safety:

'You're not hurt, Watson? For God's sake, say that you are not hurt!'

It was worth a wound—it was worth many wounds—to know the depth of loyalty and love which lay behind that cold mask. The clear, hard eyes were dimmed for a moment, and the firm lips were shaking. For the one and only time I caught a glimpse of the great heart as well as of a great brain. All my years of humble and single-minded service culminated in that moment of revelation.

At the end of *The Sign of the Four*, however, their friendship appears to be threatened when Watson tells Holmes that he is to be married, bringing an end to the cosy bachelor environment they have enjoyed together in Baker Street and depriving Holmes of possibly the only true friend he has ever had. It was, though, a traditional happy ending for Watson, and a means of emphasizing the essentially solitary and self-sufficient nature of Holmes. But when Conan Doyle came to write further exploits of Holmes he had to find a way of reuniting the pair. He projected, almost instinctively, the outstanding means of audience identification: a narrator exactly like our ordinary selves or our doctors, unusual only in his friend, on whom he may call, who may look in, or whose earlier exploits he may remember. And it was this which captured the international audience and established the cult.

The relationship between Holmes and Watson is given substance in *The Sign of the Four* by the details of their domestic life together in Baker Street. We encounter them lounging together in bachelor comfort, though Watson nurses an old leg wound, which 'ached wearily at every change of the weather', while Holmes chafes at 'the dull routine of existence'. Neither suffers from the distraction of emotional attachment. Their rooms are a haven from such disruptions—comfortable, masculine, club-like. The only female is their housekeeper Mrs Hudson (her maid is mentioned in *A Study in Scarlet*). Until Watson announces his impending marriage at the end of the book it is a world insulated from sexual intrusion—above all, the closeness that exists between them is completely purged of sexual innuendo. Ironically, it is a female, Mary Morstan, who breaks the settled peace of their masculine existence and, by doing so, provides Holmes with the kind of intellectual challenge he has been craving.

Beyond the agreeable confines of 221B Baker Street, London as a whole is a palpable and powerful presence.

It was a September evening and not yet seven o'clock, but the day had been a dreary one, and a dense drizzly fog lay low upon the great city. Mud-coloured clouds drooped sadly over the muddy streets. Down the Strand the lamps were but misty splotches of diffused light which threw a feeble circular glimmer upon the slimy pavement. The yellow glare from the shop-windows streamed out into the steamy, vaporous air and threw a murky, shifting radiance

across the crowded thoroughfare. There was, to my mind, something eerie and ghostlike in the endless procession of faces which
flitted across these narrow bars of light—sad faces and glad,
haggard and merry.

Conan Doyle was of course writing about contemporary
realities, but descriptions like this have become part of the
Sherlock Holmes myth, and images of a gaslit, fog-shrouded
London an important part of its nostalgic appeal. Holmes
demonstrates an exact knowledge of the city; but his creator's familiarity with the capital was in fact limited, as he
admitted in a letter to Stoddart on 6 March 1890: 'By the
way it must amuse you to see the vast and accurate
knowledge of London which I display. I worked it all out
from a post-office map.' Despite—or because of—this lack
of first-hand knowledge, the evocation of metropolitan
mood and atmosphere is one of the successes of *The Sign of
the Four*, even more with river than with streets.

   The speed at which *The Sign of the Four* was written meant
that Conan Doyle drew on sources immediately available to
him for background detail and had no time to confirm the
accuracy of his facts. In particular, his characterization of
Tonga was based on a misconception, which elicited a
modest censure from his former sponsor Andrew Lang in his
'The Novels of Sir Arthur Conan Doyle' *Quarterly Review*,
July 1904:

The Andamanese are cruelly libelled, and have neither the malig-
nant qualities, nor the heads like mops, nor the weapons, nor the
customs, with which they are credited by Sherlock ... if Mr
Sherlock Holmes, instead of turning up a common work of
reference, had merely glanced at the photographs of Andamanese,
trim, elegant, closely-shaven men ... he would have sought elsewhere for his little savage villain with the blow-pipe. A Fuegian who
had lived a good deal on the Amazon might have served his turn.

Another slip concerns Jonathan Small's Sikh colleagues—
Mahomet Singh, Abdullah Khan, and Dost Akbar—whose
names appear in fact to be mostly Muhammadan. ('Singh',
impossibly for linkage to 'Mahomet', is Sikh.) Such incon-

sistencies indicate that Conan Doyle's chief interest was to tell the story: he was less concerned with the verification of factual detail than with the dynamics of narrative and plot, to the extent that he never bothered to correct mistakes of fact or discrepancies in later editions. They resemble the similarly named trio of *The Mystery of Cloomber* in being murderers handled with some sympathy, especially in relation to rapacious British officers. Conan Doyle naturally championed the victims.

Conan Doyle's inventive facility begins to show itself clearly in *The Sign of the Four*. His choice of names for his characters, for instance, adds much to the story's impact (over the course of the Sherlock Holmes cycle Conan Doyle was to show himself to be almost as imaginative in this respect as Dickens): Pondicherry Lodge, Thaddeus and Bartholomew Sholto, Mordecai Smith, Athelney Jones. He was also now demonstrating a sure touch in creating an atmosphere of suspense and menace. When Bartholomew Sholto is found murdered, apparently in a locked and barred room, the exhortations of his housekeeper insinuate the possibility of some supernatural agency and by so doing intensify the melodramatic effect: ' "You must go up, Mr Thaddeus—you must go up and look for yourself. I have seen Mr Bartholomew Sholto in joy and in sorrow for ten long years, but I never saw him with such a face on him as that." ' Such descriptions are common enough in Victorian ghost stories; but here supernatural possibilities are evoked only to be dismissed by Holmes, the super-rationalist. Though Conan Doyle himself was constitutionally disposed towards an acceptance of supernatural phenomena, he was to allow his detective to remain consistently sceptical on the matter. As Holmes tells Watson in 'The Sussex Vampire' (*Case-Book*): ' "This agency stands flat-footed upon the ground, and there it must remain. The world is big enough for us. No ghosts need apply." ' (Written in 1923, when the author was Spiritualism's apostle.)

The Sign of the Four increases the small touches of humour, a feature of the ensuing cycle whose uses proclaim a rapidly developing confidence. During the chase, for example, the 'infallible' Toby, following the trail left by Small and Tonga, arrives with a triumphant yelp at a cask of creosote, at which Holmes laughs as heartily as Watson. Conan Doyle's capacity for delicate, clear, firm characterization—which he was to deploy to great effect when he turned to the medium of the short story—is apparent throughout *The Sign of the Four*. The description of Mary Morstan, which also illustrates the sympathetic care he took in delineating countenance and dress of his female characters, is typical—as is the mingling of the observant doctor and the susceptible Watson:

She was a blonde lady, small, dainty, well gloved, and dressed in the most perfect taste. There was, however, a plainness and simplicity about her costume which bore with it a suggestion of limited means. The dress was a sombre grayish beige, untrimmed and unbraided, and she wore a small turban of the same dull hue, relieved only by a suspicion of white feather in the side. Her face had neither regularity of feature nor beauty of complexion, but her expression was sweet and amiable, and her large blue eyes were singularly spiritual and sympathetic.

In structural terms *The Sign of the Four* again improves on *A Study in Scarlet*, whose lengthy flashback disrupts the flow of the story. Conan Doyle was never completely at home when relating the exploits of Holmes in the novel format: he was far more successful, and conspicuously at his ease, in the short stories that began in the *Strand Magazine* in July 1891. But *The Sign of the Four* is the most coherent of the long narratives—including the more famous *The Hound of the Baskervilles* (1902), in which Holmes is absent from the action for much of the time.

The main line of the narrative, the adventure itself, alternates with Watson's courtship of, and eventual engagement to, Mary Morstan. The love interest develops from their first meeting: 'In an experience of women which extends over many nations and three separate continents, I have never looked upon a face which gave a clearer promise of a refined and sensitive nature.' By the end of the chapter

Watson is also musing on the possibility of a relationship with Holmes's client:

So I sat and mused until such dangerous thoughts came into my head that I hurried away to my desk and plunged furiously into the latest treatise upon pathology. What was I, an army surgeon with a weak leg and a weaker banking account, that I should dare to think of such things? She was a unit, a factor—nothing more. If my future were black, it was better surely to face it like a man than to attempt to brighten it by mere will-o'-the-wisps of the imagination.

But whilst travelling by cab to meet Thaddeus Sholto, Watson continues the courtesies to Mary Morstan, leaving Holmes, deep in thought, to consider the problem in hand. Once in the garden of Pondicherry Lodge, while Holmes surveys the devastation caused by the Sholtos' excavations,

Miss Morstan and I stood together, and her hand was in mine. A wondrous subtle thing is love, for here were we two, who had never seen each other before that day, between whom no word or even look of affection had ever passed, and yet now in an hour of trouble our hands instinctively sought for each other.

As the plot unfolds, Watson openly questions his motives and their possible interpretation: 'Might she not look upon me as a mere vulgar fortune-seeker? I could not bear to risk that such a thought should cross her mind.'

Watson was present when Mary Morstan opened the treasure chest recently retrieved from Jonathan Small. The relief expressed when it is realized that the treasure is missing, is almost tangible. Watson realizes that, at last, Mary is within his reach and his love for her need no longer be suppressed:

'The treasure is lost', said Miss Morstan, calmly.
As I listened to the words and realized what they meant, a great shadow seemed to pass from my soul. I did not know how this Agra treasure had weighed me down, until now that it was finally removed. It was selfish, no doubt, disloyal, wrong, but I could realize nothing save that the golden barrier was gone from between us.
'Thank God!' I ejaculated from my very heart.

She looked at me with a quick, questioning smile.

'Why do you say that?' she asked.

'Because you are within my reach again', I said, taking her hand. She did not withdraw it. 'Because I love you, Mary, as truly as ever a man loved a woman. Because this treasure, these riches, sealed my lips. Now that they are gone I can tell you how I love you. That is why I said, "Thank God." '

'Then I say "Thank God", too,' she whispered, as I drew her to my side.

Whoever had lost a treasure, I knew that night that I had gained one.

Though *Lippincott's Magazine* had commissioned the story to be published in a single issue, the love episodes in *The Sign of the Four* occur at the beginnings and ends of chapters, almost as if Conan Doyle was writing for serialization. That he succeeded in introducing this second element to the plot without its being overshadowed by the main drama of the quest and its outcome was due to the strength of the characters involved. Watson is the story's narrative voice: we see through his eyes, share his thoughts, and his individuality is continually conveyed. Miss Morstan, too, stands clear before us. She is far more than a passive damsel in distress, as Holmes concedes on learning of Watson's engagement: ' "She is one of the most charming young ladies I ever met and might have been most useful in such work as we have been doing. She had a decided genius that way" '. Behind both the protagonists in the love plot are autobiographical resonances. The tenderness and strength of the love passages seem to hint at bases in Conan Doyle and his first wife, Louisa. Very little is known of their courtship and love, and this may be the clearest glimpse of it we will ever have. Like Louisa Hawkins, Mary Morstan is blonde, blue-eyed, and 27 years of age (Louisa's age when she married Conan Doyle).

For the character of Holmes himself there are several probable influences, principally the surgeon Joseph Bell (1837–1911) under whom Conan Doyle studied as a medical student. In *The Sign of the Four* Bell is recalled in the passage in which Holmes deduces, from the reddish mould adhering to his instep, that Watson had visited the Wigmore Street Post Office: this is borrowed from the occasion when an Irish patient of Bell's arrived with reddish clay on his boots that the doctor identified as coming from the Bruntsfield Links. In his *Manual of the Operations of Surgery* (1883), Bell acknowledges Dr Patrick Heron Watson for his plaster-of-Paris moulds; in *The Sign of the Four* Sherlock Holmes is similarly proficient in the use of the same substance: ' "Here is my monograph upon the tracing of footsteps, with some remarks upon the uses of plaster of Paris as a preserver of impresses." ' (Owen Dudley Edwards[4] has pointed out the proximity in Bell's *Manual* between the names of Mr [Timothy] Holmes and Dr [Patrick Heron] Watson.)

One other person from Conan Doyle's early days in medical practice who probably contributed to Holmes's character was Dr George Turnavine Budd (1855–1889). Conan Doyle's relationship with Budd, the model for Cullingworth in *The Stark Munro Letters* (1895), seems to have provided a good deal of enjoyment, but it was also a turbulent one which ended in bitterness and recrimination after an ill-fated partnership in Plymouth early in 1882. In *The Stark Munro Letters* an evening's conversation is recalled in which Cullingworth puts forward his ideas for revolutionizing the future of warships: it cannot be coincidental that this is one of the topics on which Holmes discourses:

Holmes could talk exceedingly well when he chose, and that night he did choose. He appeared to be in a state of nervous exaltation. I have never known him so brilliant. He spoke on a quick succession of subjects—on miracle plays, on medieval pottery, on Stradivarius violins, on the Buddhism of Ceylon, and on the warships of the future—handling each as though he had made a special study of it.

The other topics (apart from Strads) seem more like an evening's entertainment with Major-General Drayson.

---

[4] O. D. Edwards, *The Quest for Sherlock Holmes* (Edinburgh: Mainstream, 1983), 205.

Conan Doyle was only 30 when he wrote *The Sign of the Four* but already he was displaying the qualities that were to turn him into one of the most popular writers of his day. Though his literary influences were still apparent, he had matured rapidly as a writer since the publication of *A Study in Scarlet*. The plot of the new Holmes novel, though derivative in places (much less so than its predecessor), was confidently worked out and conveyed through well-paced narrative and crisp dialogue. Above all, the central character of Sherlock Holmes had an intensity of realization that gave the book coherence and mesmerizing interest. It also —crucially—offered the prospect of greater things to come.

The Sign of the Four had moved Holmes from *A Study in Scarlet*'s retrospective Watson looking back through otherwise unidentified Reminiscences, to the contemporary Watson telling of recent events. To do this it had been necessary to suggest an intimate yet brief sojourn for both men in Baker Street: they know each other well, yet the first chapter prompts revelations improbable in an extended co-tenancy. By sleight of hand Watson is made to have seen much of Holmes in action, while specifically no case seems to have made an impact on him save that which he entitled *A Study in Scarlet*. In part, this is a device to turn the sequel of an old story into yesterday's tale and give it purchase on tomorrow; in part, it is an increasingly professional writer forcing the earlier work on his new audience. Now they are ready for a series of self-standing short stories, complete in themselves but making the occasional allusion to previous cases. It would be some years before Conan Doyle had acquired sufficient confidence to invent untold cases whose titles were tantalizing gems. But the ground was now cleared for the *Adventures*, to be begun fourteen months after publication of *The Sign of the Four*.

The *Adventures* might well be an inevitability on the basis of *The Sign of the Four*, but Holmes's iconographic future was as uncertain as ever. And then, in the unexpected hands of Sidney Paget (1860–1908) the drawings for the *Adventures* memorably milestoned the *Strand* road to Holmes's immortality. One could only say that *The Sign of the Four* found better artists than the *A Study in Scarlet. Lippincott's* gave *The Sign of the Four* one picture—alas, without Holmes or Watson, though much the finest a Holmes story had yet received. It faced the Agra part of Jonathan Small's 'strange story', captioned 'I shall reward you, young Sahib, and your governor also, if he will give me the shelter I ask.' The pseudo-merchant Achmet, small, cherubic, turbaned, baby-faced, all trust and innocence, looks into the half-visible profiled face of the young and even more innocent Jonathan Small, wooden-legged, legionnaire-headdressed, honesty and compassion making their last stand within him under the anxious eyes of the Sikh guards out of Achmet's sight, while behind Achmet the huge-bearded Dost Akbar, thin, judicial, the only one in total control, awaits the future with grim certainty. It is Small's moment of choice between life and honour, and the humanity so fully realized in all the participants makes it the more heartrending. It is the work of Herbert Denman (1855–1903), Paris Salon prizewinner in 1886, Paris Expo prizewinner in 1889, future designer of the Waldorf-Astoria ballroom in New York.

The Sign of the Four was reprinted as such in the Bristol *Observer*, weekly, from 17 May to 5 July 1890, the first mass circulation Holmes had so far achieved, with illustrations giving moustaches to both Holmes and Watson, and (for the first time ever) a deerstalker to Holmes. (Weekly serializations to local newspaper audiences also ran in the *Hampshire Telegraph and Sussex Chronicle*, July–August, and the *Birmingham Weekly Mercury* August–September 1890, old home grounds of the author and arranged by him.) Charles Kerr (1858–1907) in Spencer Blackett's edition retained a moustache for its sadistic, threatening-looking Holmes who smirks like a Lestrade with an eerie foretaste of Basil Rathbone at his most unpleasant. Captioned 'In the light of the lantern I read, with a thrill of horror, "the sign of the four" ', Watson seems to be screaming his head off while the corpse of Bartholomew Sholto enjoys the general discomfiture with a sociable leer. Neither these nor the disastrous attempts for *A Study in Scarlet* put constraint on the future work of Sidney Paget. Nor were his laurels at any real risk from the George Newnes 'Souvenir' edition of 1903 when eight illustrations were added by Frederick Henry Townsend (1868–1920), appointed Art Editor of *Punch* in 1905: the artist had drawn Sir Henry Irving (1838–1905) in the *Pall Mall Budget* for 27 September 1894 playing the aged corporal in Conan Doyle's play *Waterloo*, but while his Holmes in disguise as the old man of Chapter IX made appropriate efforts along the same direction, his physiognomy *en clair* suggested a gentlemanly sports master. There was justifiable contemporary admiration for Townsend's frontispiece 'death of Bartholomew Sholto' with its Cheshire-cat Tonga on the ladder while the corpse raises an indignant half-clenched left hand to Heaven for vengeance.

But Tonga was the central figure in perhaps the finest of all illustrations for *The Sign of the Four*, by one of Conan Doyle's favourites among the artists of the *Strand* after Sidney Paget's death, Arthur Twidle (1865–1936): the location was in the Smith, Elder 'Author's Edition' of Conan Doyle's collected works (1903). Captioned 'He whirled round, and fell sideways into the stream', it has a gaunt, athletic Small throwing himself on the tiller while Tonga, blowpipe clenched aloft, but drawing his legs up in convulsive pain—reflected in a face and naked breast of fierce, youthful nobility—is thrown to his doom; flashes from the pursuing craft and flames from its funnel shade Holmes's implacable face, firing, with Watson, who would seem to have fired first, a silhouette at his side.

Strangely enough, the only Holmes artist who even remotely rivalled Sidney Paget in his lifetime, the American Frederic Dorr Steele (1873–1944), who first entered the field with his illustrations for the serialized *Return of Sherlock Holmes* in *Collier's Weekly* in 1903, was brazenly re-used for a pirated edition of *The Sign of the Four* in possibly the most extraordinary case of iconographic recycling modern literature affords. A three-volume pirate collection *Conan Doyle's Best Books*, published in 1904 by P. F. Collier of New York and much prized by collectors for its inclusion of many fugitive juvenilia, took several Steele pictures from the *Return* and captioned them as illustrations for *The Sign of the Four*, with Dr Thorneycroft Huxtable ('The Priory School', *Return*) refreshing himself with milk and biscuits becoming 'Major John Sholte [sic], once of the Indian Army', while Holmes after his knocking-out of Woodley in the pub ('The Solitary Cyclist', *Return*) becomes Holmes reprimanding Athelney Jones for premature conclusions on first looking into Bartholemew's body. Occasionally, Conan Doyle re-worked an old idea or considered a road he had not previously taken in a comparable plot; but on Collier's showing, every Holmes story was another *Sign of the Four*. Irrespective of this rapacious lunacy, one can see its long shadows in many places, and 'The Crooked Man' (*Memoirs*) in particular suggests a sympathy with Jonathan Small (as against the officer class) needing to find an outlet. But *The Sign of the Four* built up its own enormous following once it became a George Newnes title alongside the *Adventures* and *Memoirs* in 1893, and seems to have outsold its fellows, just as in the United States its piracies dwarf the number of almost all other Holmes titles and those of virtually any other author of its day. One has to go back to the primes of Charles Dickens, Walter Scott, and Thomas Babington Macaulay for anything comparable.[5]

As critics need reminding, the reviews (save for their vital role in winning separate publication for *A Study in Scarlet*) played little part in the success of the Holmes series, but two notices of the book publication of *The Sign of the Four* merit reproduction in their own right. George Cotterell welcomed it in the *Academy* for 13 December 1890: an Edinburgh political satirist and poet, Cotterell may have had some connection with the *Scotsman* circle which had applauded *A Study in Scarlet* for all of its *Beeton's Christmas Annual* ephemeral

---

[5] For full discussion of the book's illicit history, the reader will be well rewarded by Donald A. Redmond's *Sherlock Holmes Among the Pirates—Copyright and Conan Doyle in America 1890–1930* (1990).

packaging, and his own publisher was William Blackwood (1836–1912), whose magazine had published Conan Doyle's 'A Physiologist's Wife' in its September number. So as with Conan Doyle's other early work, a friendly Scots hand eased the way onward:

Detective stories always have a certain charm, and perhaps the charm is greatest when the detective element is non-professional. The accomplished amateur in the fine art of discovering crime and hunting down the criminal is a much more wonderful personage than the official detective. At any rate, Sherlock Holmes, in *The Sign of Four*, was such a personage. The curious incidents, the mystery of which he unravels, make a capital story, which is told with a directness that keeps the reader's attention fixed till he gets to the sequel. After the sequel, as part of the story, follows the narrative of the man who has been hunted down; and though this is interesting in itself, and has a bearing upon the plot, it is somewhat flat after a breathless chase which has been breathlessly described. It has the effect of an anti-climax. Sherlock Holmes is the best-drawn of the characters, perhaps because there was most character in him to draw. The young lady is rather insipid, but she had not much to say or to do. The man with a wooden leg, who was nearly a match for Holmes, is also nearest him in point of vivid portraiture.

The *Glasgow Herald*, too, had wallowed in the word 'wonderful' for Holmes in reviewing *A Study in Scarlet*, although its writer was more starry-eyed than the hard-bitten Cotterell. But the most telling feature of Cotterell's review is surely its author's unconscious glide into commentary as though Holmes, Mary Morstan, and Small were real: as criticism, it was more astute than its author realized; as fantasy, it was the first drop in an ocean.

The anonymous *Athenaeum* reviewer on 6 December 1890, unlike Cotterell, has been reprinted hitherto,[6] but it well merits further exhumation:

A detective story is usually lively reading, but we cannot pretend to think that *The Sign of Four* is up to the level of the writer's best

---

[6] In Howard Haycraft (ed.), *The Art of the Mystery Story* (1947), 381.

work. It is a curious medley, and full of horrors; and surely those who play hide and seek with the fatal treasure are a curious company. The wooden-legged convict and his fiendish misshapen little mate, the ghastly twins, the genial prizefighters, the detectives wise and foolish, and the gentle girl whose lover tells the tale, twist in and out together in a mazy dance, culminating in that mad and terrible rush down the river which ends mystery and the treasure. Dr Doyle's admirers will read the little volume through eagerly enough, but they will hardly care to take it up again.

The prophesy has surely been conclusively confounded. Graham Greene—at the age of 70—spoke for many: '*The Sign of Four* . . . I read first at the age of ten and have never forgotten . . . the dark night in Pondicherry Lodge, Norwood, has never faded from my memory.' We hardly need further testimony to the power and the glory of *The Sign of the Four*.

CHRISTOPHER RODEN

NOTE ON THE TEXT

The text of this edition is based on the third edition of the book version (1893) published by George Newnes. This has been collated with the first published text (*Lippincott's Magazine*, Feb. 1890), the Author's Edition (1903), published by Smith, Elder, and the Doubleday *Complete Sherlock Holmes* (1930). Notice has also been taken of other editions.

SELECT BIBLIOGRAPHY

I. A. CONAN DOYLE: PRINCIPAL WORKS

(a) Fiction

A Study in Scarlet (Ward, Lock, & Co., 1888)

The Mystery of Cloomber (Ward & Downey, 1888)

Micah Clarke (Longmans, Green, & Co., 1889)

The Captain of the Pole-Star and Other Tales (Longmans, Green, & Co., 1890)

The Sign of the Four (Spencer Blackett, 1890)

The Firm of Girdlestone (Chatto & Windus, 1890)

The White Company (Smith, Elder, & Co., 1891)

The Adventures of Sherlock Holmes (George Newnes, 1892)

The Great Shadow (Arrowsmith, 1892)

The Refugees (Longmans, Green, & Co., 1893)

The Memoirs of Sherlock Holmes (George Newnes, 1893)

Round the Red Lamp (Methuen & Co., 1894)

The Stark Munro Letters (Longmans, Green, & Co., 1895)

The Exploits of Brigadier Gerard (George Newnes, 1896)

Rodney Stone (Smith, Elder, & Co., 1896)

Uncle Bernac (Smith, Elder, & Co., 1897)

The Tragedy of the Korosko (Smith, Elder, & Co., 1898)

A Duet With an Occasional Chorus (Grant Richards, 1899)

The Green Flag and Other Stories of War and Sport (Smith, Elder, & Co., 1900)

The Hound of the Baskervilles (George Newnes, 1902)

Adventures of Gerard (George Newnes, 1903)

The Return of Sherlock Holmes (George Newnes, 1905)

Sir Nigel (Smith, Elder, & Co., 1906)

Round the Fire Stories (Smith, Elder, & Co., 1908)

The Last Galley (Smith, Elder, & Co., 1911)

The Lost World (Hodder & Stoughton, 1912)

The Poison Belt (Hodder & Stoughton, 1913)

The Valley of Fear (Smith, Elder, & Co., 1915)

His Last Bow (John Murray, 1917)

Danger! and Other Stories (John Murray, 1918)

The Land of Mist (Hutchinson & Co., 1926)

The Case-Book of Sherlock Holmes (John Murray, 1927)

The Maracot Deep and Other Stories (John Murray, 1929)

The Complete Sherlock Holmes Short Stories (John Murray, 1928)
The Conan Doyle Stories (John Murray, 1929)
The Complete Sherlock Holmes Long Stories (John Murray, 1929)

(b) *Non-fiction*

The Great Boer War (Smith, Elder, & Co., 1900)
The Story of Mr George Edalji (T. Harrison Roberts, 1907)
Through the Magic Door (Smith, Elder, & Co., 1907)
The Crime of the Congo (Hutchinson & Co., 1909)
The Case of Oscar Slater (Hodder & Stoughton, 1912)
The German War (Hodder & Stoughton, 1914)
The British Campaign in France and Flanders (Hodder & Stoughton, 6 vols., 1916–20)
The Poems of Arthur Conan Doyle (John Murray, 1922)
Memories and Adventures (Hodder & Stoughton, 1924; revised edn., 1930)
The History of Spiritualism (Cassell & Co., 1926)

2. MISCELLANEOUS

A Bibliography of A. Conan Doyle (Soho Bibliographies 23: Oxford, 1983) by Richard Lancelyn Green and John Michael Gibson, with a foreword by Graham Greene, is the standard—and indispensable—source of bibliographical information, and of much else besides. Green and Gibson have also assembled and introduced *The Unknown Conan Doyle*, comprising *Uncollected Stories* (those never previously published in book form); *Essays in Photography* (documenting a little-known enthusiasm of Conan Doyle's during his time as a student and young doctor), both published in 1982; and *Letters to the Press* (1986). Alone, Richard Lancelyn Green has compiled (1) *The Uncollected Sherlock Holmes* (1983), an impressive assemblage of Holmesiana, containing almost all Conan Doyle's writing about his creation (other than the stories themselves) together with related material by Joseph Bell, J. M. Barrie, and Beverley Nichols; (2) *The Further Adventures of Sherlock Holmes* (1985), a selection of eleven apocryphal Holmes adventures by various authors, all diplomatically introduced; (3) *The Sherlock Holmes Letters* (1986), a collection of noteworthy public correspondence on Holmes and Holmesiana and far more valuable than its title suggests; and (4) *Letters to Sherlock Holmes* (1984), a powerful testimony to the power of the Holmes stories.

Though much of Conan Doyle's work is now readily available there are still gaps. Some of his very earliest fiction now only survives in rare piracies (apart, that is, from the magazines in which they were first published), including items of intrinsic genre interest such as 'The Gully of Bluemansdyke' (1881) and its sequel 'My Friend the Murderer' (1882), which both turn on the theme of the murderer-informer (handled very differently—and far better—in the Holmes story of 'The Resident Patient' (*Memoirs*)): both of these were used as book-titles for the same pirate collection first issued as *Mysteries and Adventures* (1889). Other stories achieved book publication only after severe pruning—for example, 'The Surgeon of Gaster Fell', reprinted in *Danger!* many years after magazine publication (1890). Some items given initial book publication were not included in the collected edition of *The Conan Doyle Stories*. Particularly deplorable losses were 'John Barrington Cowles' (1884: included subsequently in *Edinburgh Stories of Arthur Conan Doyle* (1981)), 'A Foreign Office Romance' (1894), 'The Club-Footed Grocer' (1898), 'A Shadow Before' (1898), and 'Danger!' (1914). Three of these may have been post-war casualties, as seeming to deal too lightheartedly with the outbreak of other wars; 'John Barrington Cowles' may have been dismissed as juvenile work; but why Conan Doyle discarded a story as good as 'The Club-Footed Grocer' would baffle even Holmes.

At the other end of his life, Conan Doyle's tidying impaired the survival of his most recent work, some of which well merited lasting recognition. *The Maracot Deep and Other Stories* appeared in 1929, a little over a month after *The Conan Doyle Stories*; 'Maracot' itself found a separate paperback life as a short novel; the two Professor Challenger stories, 'The Disintegration Machine' and 'When the World Screamed', were naturally included in John Murray's *The Professor Challenger Stories* (1952); but the fourth item, 'The Story of Spedegue's Dropper', passed beyond the ken of most of Conan Doyle's readers. These three stories show the author, in his seventieth year, still at the height of his powers.

In 1980 Gaslight Publications, of Bloomington, Ind., reprinted *The Mystery of Cloomber*, *The Firm of Girdlestone*, *The Doings of Raffles Haw* (1892), *Beyond the City* (1893), *The Parasite* (1894; also reprinted in *Edinburgh Stories of Arthur Conan Doyle*), *The Stark Munro Letters*, *The Tragedy of the Korosko*, and *A Duet*. *Memories and Adventures*, Conan Doyle's enthralling but impressionistic recollections, are best read in the revised (1930) edition. *Through the Magic Door* remains the best introduction to the literary mind of Conan Doyle, whilst some of his volumes on Spiritualism have autobiographical material of literary significance.

ACD: The Journal of the Arthur Conan Doyle Society (ed. Christopher Roden, David Stuart Davies [to 1991], and Barbara Roden [from 1992]), together with its newsletter, *The Parish Magazine*, is a useful source of critical and biographical material on Conan Doyle. The enormous body of 'Sherlockiana' is best pursued in *The Baker Street Journal*, published by Fordham University Press, or in the *Sherlock Holmes Journal* (Sherlock Holmes Society of London), itemized up to 1974 in the colossal *World Bibliography of Sherlock Holmes and Doctor Watson* (1974) by Ronald Burt De Waal (see also De Waal, *The International Sherlock Holmes* (1980)) and digested in *The Annotated Sherlock Holmes* (2 vols., 1968) by William S. Baring-Gould, whose industry has been invaluable for the Oxford Sherlock Holmes editors. Jack Tracy, *The Encyclopaedia Sherlockiana* (1979) is a very helpful compilation of relevant data. Those who can nerve themselves to consult it despite its title will benefit greatly from Christopher Redmond, *In Bed With Sherlock Holmes* (1984). The classic 'Sherlockian' work is Ronald A. Knox, 'Studies in the Literature of Sherlock Holmes', first published in *The Blue Book* (July 1912) and reprinted in his *Essays in Satire* (1928).

The serious student of Conan Doyle may perhaps deplore the vast extent of 'Sherlockian' literature, even though the size of this output is testimony in itself to the scale and nature of Conan Doyle's achievement. But there is undoubtedly some wheat amongst the chaff. At the head stands Dorothy L. Sayers, *Unpopular Opinions* (1946); also of some interest are T. S. Blakeney, *Sherlock Holmes: Fact or Fiction* (1932), H. W. Bell, *Sherlock Holmes and Dr Watson* (1932), Vincent Starrett, *The Private Life of Sherlock Holmes* (1934), Gavin Brend, *My Dear Holmes* (1951), S. C. Roberts, *Holmes and Watson* (1953) and Roberts's introduction to *Sherlock Holmes: Selected Stories* (Oxford: The World's Classics, 1951), James E. Holroyd, *Baker Street Byways* (1959), Ian McQueen, *Sherlock Holmes Detected* (1974), and Trevor H. Hall, *Sherlock Holmes and his Creator* (1978). One Sherlockian item certainly falls into the category of the genuinely essential: D. Martin Dakin, *A Sherlock Holmes Commentary* (1972), to which all the editors of the present series are indebted.

Michael Pointer, *The Public Life of Sherlock Holmes* (1975) contains invaluable information concerning dramatizations of the Sherlock Holmes stories for radio, stage, and the cinema; of complementary interest are Chris Steinbrunner and Norman Michaels, *The Films of Sherlock Holmes* (1978) and David Stuart Davies, *Holmes of the Movies* (1976), whilst Philip Weller with Christopher Roden, *The Life and Times of Sherlock Holmes* (1992) summarizes a great deal of useful information concerning Conan Doyle's life and Holmes's cases, and in addition is delightfully illustrated. The more concrete products of the Holmes industry are dealt with in Charles Hall, *The Sherlock Holmes Collection* (1987). For a useful retrospective view, Allen Eyles, *Sherlock Holmes: A Centenary Celebration* (1986) rises to the occasion. Both useful and engaging are Peter Haining, *The Sherlock Holmes Scrapbook* (1973) and Charles Viney, *Sherlock Holmes in London* (1989).

Of the many anthologies of Holmesiana, P. A. Shreffler (ed.), *The Baker Street Reader* (1984) is exceptionally useful. D. A. Redmond, *Sherlock Holmes: A Study in Sources* (1982) is similarly indispensable. Michael Hardwick, *The Complete Guide to Sherlock Holmes* (1986) is both reliable and entertaining; Michael Harrison, *In the Footsteps of Sherlock Holmes* (1958) is occasionally helpful.

For more general studies of the detective story, the standard history is Julian Symons, *Bloody Murder* (1972, 1985, 1992). Necessary but a great deal less satisfactory is Howard Haycraft, *Murder for Pleasure* (1942); of more value is Haycraft's critical anthology *The Art of the Mystery Story* (1946), which contains many choice period items. Both R. F. Stewart, *... And Always a Detective* (1980) and Colin Watson, *Snobbery with Violence* (1971) are occasionally useful. Dorothy Sayers's pioneering introduction to *Great Short Stories of Detection, Mystery and Horror* (First Series, 1928), despite some inspired howlers, is essential reading; Raymond Chandler's riposte, 'The Simple Art of Murder' (1944), is reprinted in Haycraft, *The Art of the Mystery Story* (see above). Less well known than Sayers's essay but with an equal claim to poineer status is E. M. Wrong's introduction to *Crime and Detection*, First Series (Oxford: The World's Classics, 1926). See also Michael Cox (ed.), *Victorian Tales of Mystery and Detection: An Oxford Anthology* (1992).

Amongst biographical studies of Conan Doyle one of the most distinguished is Jon L. Lellenberg's survey, *The Quest for Sir Arthur Conan Doyle* (1987), with a Foreword by Dame Jean Conan Doyle (much the best piece of writing on ACD by any member of his family). The four earliest biographers—the Revd John Lamond (1931), Hesketh Pearson (1943), John Dickson Carr (1949), and Pierre Nordon (1964)—all had access to the family archives, subsequently closed to researchers following a lawsuit; hence all four biographies contain valuable documentary material, though Nordon handles the evidence best (the French text is fuller than the English version, published in 1966). Of the others, Lamond seems only to have made little use of the material available to him;

Pearson is irreverent and wildly careless with dates; Dickson Carr has a strong fictionalizing element. Both he and Nordon paid a price for their access to the Conan Doyle papers by deferring to the far from impartial editorial demands of Adrian Conan Doyle; Nordon nevertheless remains the best available biography. The best short sketch is Julian Symons, *Conan Doyle* (1979) (and for the late Victorian milieu of the Holmes cycle some of Symons's own fiction, such as *The Blackheath Poisonings* and *The Detling Secret*, can be thoroughly recommended). Harold Orel (ed.), *Critical Essays on Sir Arthur Conan Doyle* (1992) is a good and varied collection, whilst Robin Winks, *The Historian as Detective* (1969) contains many insights and examples applicable to the Holmes corpus; Winks's *Detective Fiction: A Collection of Critical Essays* (1980) is an admirable working handbook, with a useful critical bibliography. Edmund Wilson's famous essay 'Mr Holmes, they were the footprints of a gigantic hound' (1944) may be found in his *Classics and Commercials: A Literary Chronicle of the Forties* (1950).

Specialized biographical areas are covered in Owen Dudley Edwards, *The Quest for Sherlock Holmes: A Biographical Study of Arthur Conan Doyle* (1982) and in Geoffrey Stavert, *A Study in Southsea: The Unrevealed Life of Dr Arthur Conan Doyle* (1987), which respectively assess the significance of the years up to 1882, and from 1882 to 1890. Alvin E. Rodin and Jack D. Key provide a thorough study of Conan Doyle's medical career and its literary implications in *Medical Casebook of Dr Arthur Conan Doyle* (1984). Peter Costello, in *The Real World of Sherlock Holmes: The True Crimes Investigated by Arthur Conan Doyle* (1991) claims too much, but it is useful to be reminded of events that came within Conan Doyle's orbit, even if they are sometimes tangential or even irrelevant. Christopher Redmond, *Welcome to America, Mr Sherlock Holmes* (1987) is a thorough account of Conan Doyle's tour of North America in 1894.

Other than Baring-Gould (see above), the only serious attempt to annotate the nine volumes of the Holmes cycle has been in the Longman Heritage of Literature series (1979–80), to which the present editors are also indebted. Of introductions to individual texts, H. R. F. Keating's to the *Adventures* and *The Hound of the Baskervilles* (published in one volume under the dubious title *The Best of Sherlock Holmes* (1992)) is worthy of particular mention.

A CHRONOLOGY OF ARTHUR CONAN DOYLE

1855   Charles Altamont Doyle, youngest son of the political cartoonist John Doyle ('HB'), and Mary Foley, his Irish landlady's daughter, marry in Edinburgh on 31 July.

1859   Arthur Ignatius Conan Doyle, third child and elder son of ten siblings, born at 11 Picardy Place, Edinburgh, on 22 May and baptized into the Roman Catholic religion of his parents.

1868–75 ACD commences two years' education under the Jesuits at Hodder, followed by five years at its senior sister college, Stonyhurst, both in the Ribble Valley, Lancashire; becomes a popular storyteller amongst his fellow-pupils, writes verses, edits a school paper, and makes one close friend, James Ryan of Glasgow and Ceylon. Doyle family resides at 3 Sciennes Hill Place, Edinburgh.

1875–6   ACD passes London Matriculation Examination at Stonyhurst and studies for a year in the Jesuit college at Feldkirch, Austria.

1876–7   ACD becomes a student of medicine at Edinburgh University on the advice of Bryan Charles Waller, now lodging with the Doyle family at 2 Argyle Park Terrace.

1877–80 Waller leases 23 George Square, Edinburgh as a 'consulting pathologist', with all the Doyles as residents. ACD continues medical studies, becoming surgeon's clerk to Joseph Bell at Edinburgh; also takes temporary medical assistant-ships at Sheffield, Ruyton (Salop), and Birmingham, the last leading to a close friendship with his employer's family, the Hoares. First story published, 'The Mystery of Sasassa Valley', in *Chambers's Journal* (6 Sept. 1879); first non-fiction published—'Gelseminum as a Poison', *British Medical Journal* (20 Sept. 1879). Sometime previously ACD sends 'The Haunted Grange of Goresthorpe' to *Blackwood's Edinburgh Magazine*, but it is filed and forgotten.

1880   (Feb.–Sept.) ACD serves as surgeon on the Greenland whaler *Hope* of Peterhead.

1881   ACD graduates MB, CM (Edin.); Waller and the Doyles living at 15 Lonsdale Terrace, Edinburgh.

1881–2   (Oct.–Jan.) ACD serves as surgeon on the steamer *Mayumba* to West Africa, spending three days with US Minister to Liberia, Henry Highland Garnet, black abolitionist leader, then dying. (July–Aug.) Visits Foley relatives in Lismore, Co. Waterford.

1882   Ill-fated partnership with George Turnavine Budd in Plymouth. ACD moves to Southsea, Portsmouth, in June. ACD published in *London Society*, *All the Year Round*, *Lancet*, and *British Journal of Photography*. Over the next eight years ACD becomes an increasingly successful general practitioner at Southsea.

1882–3   Breakup of the Doyle family in Edinburgh. Charles Altamont Doyle henceforth confined because of alcoholism and epilepsy. Mary Foley Doyle resident in Masongill Cottage on the Waller estate at Masongill, Yorkshire. Innes Doyle (b. 1873) resident with ACD as schoolboy and surgery page from Sept. 1882.

1883   'The Captain of the *Pole-Star*' published (*Temple Bar*, Jan.), as well as a steady stream of minor pieces. Works on *The Mystery of Cloomber*.

1884   ACD publishes 'J. Habakuk Jephson's Statement' (*Cornhill Magazine*, Jan.), 'The Heiress of Glenmahowley' (*Temple Bar*, Jan.), 'The Cabman's Story' (*Cassell's Saturday Journal*, May); working on *The Firm of Girdlestone*.

1885   Publishes 'The Man from Archangel' (*London Society*, Jan.). Jack Hawkins, briefly a resident patient with ACD, dies of cerebral meningitis. Louisa Hawkins, his sister, marries ACD. (Aug.) Travels in Ireland for honeymoon. Awarded Edinburgh MD.

1886   Writing *A Study in Scarlet*.

1887   *A Study in Scarlet* published in *Beeton's Christmas Annual*.

1888   (July) First book edition of *A Study in Scarlet* published by Ward, Lock; (Dec.) *The Mystery of Cloomber* published.

1889   (Feb.) *Micah Clarke* (ACD's novel of the Monmouth Rebellion of 1685) published. Mary Louise Conan Doyle, ACD's eldest child, born. Unauthorized publication of *Mysteries and Adventures* (published later as *The*

Gully of Bluemansdyke and *My Friend the Murderer*). *The Sign of the Four* and Oscar Wilde's *The Picture of Dorian Gray* commissioned by Lippincott's.

1890   (Jan.) 'Mr [R. L.] Stevenson's Methods in Fiction' published in the *National Review*. (Feb.) *The Sign of the Four* published in *Lippincott's Monthly Magazine*; (Mar.) first authorized short-story collection, *The Captain of the Polestar and other tales*, published; (Apr.) *The Firm of Girdlestone* published; (Oct.) first book edition of the *Sign* published by Spencer Blackett.

1891   ACD sets up as an eye specialist in 2 Upper Wimpole Street, off Harley Street, while living at Montague Place. Moves to South Norwood. (July–Dec.) The first six 'Adventures of Sherlock Holmes' published in George Newnes's *Strand Magazine*. (Oct.) *The White Company* published; *Beyond the City* first published in *Good Cheer*, the special Christmas number of *Good Words*.

1892   (Jan.–June) Six more Holmes stories published in the *Strand*, with another in Dec. (Mar.) *The Doings of Raffles Haw* published (first serialized in Alfred Harmsworth's penny paper *Answers*, Dec. 1891–Feb. 1892). (14 Oct.) *The Adventures of Sherlock Holmes* published by Newnes. (31 Oct.) Waterloo story *The Great Shadow* published. Alleyne Kingsley Conan Doyle born. Newnes republishes the *Sign*.

1893   'Adventures of Sherlock Holmes' (second series) continues in the *Strand*, to be published by Newnes as *The Memoirs of Sherlock Holmes* (Dec.), minus 'The Cardboard Box'. Holmes apparently killed in 'The Final Problem' (Dec.) to free ACD for 'more serious literary work'. (May) *The Refugees* published. *Jane Annie; or, the Good Conduct Prize* (musical comedy co-written with J. M. Barrie) fails at the Savoy Theatre. (10 Oct.) Charles Altamont Doyle dies.

1894   (Oct.) *Round the Red Lamp*, a collection of medical short stories, published, several for the first time. *The Stark Munro Letters*, a fictionalized autobiography, begun, to be concluded the following year. ACD on US lecture tour with Innes Doyle. (Dec.) *The Parasite* published; 'The Medal of Brigadier Gerard' published in the *Strand*.

1895   'The Exploits of Brigadier Gerard' published in the *Strand*.

1896   (Feb.) *The Exploits of Brigadier Gerard* published by Newnes. ACD settles at Hindhead, Surrey, to minimize effects of his wife's tuberculosis. (Nov.) *Rodney Stone*, a pre-Regency mystery, published. Self-pastiche, 'The Field Bazaar', appears in the Edinburgh University *Student* (20 Nov.).

1897   (May) Napoleonic novel *Uncle Bernac* published; three 'Captain Sharkey' pirate stories published in *Pearson's Magazine* (Jan., Mar., May). Home at Undershaw, Hindhead.

1898   (Feb.) *The Tragedy of the Korosko* published. (June) Publishes *Songs of Action*, a verse collection. (June–Dec.) Begins to publish 'Round the Fire Stories' in the *Strand*—'The Beetle Hunter', 'The Man with the Watches', 'The Lost Special', 'The Sealed Room', 'The Black Doctor', 'The Club-Footed Grocer', and 'The Brazilian Cat'. Ernest William Hornung (ACD's brother-in-law) creates A. J. Raffles and in 1899 dedicates the first stories to ACD.

1899   (Jan.–May) Concludes 'Round the Fire' series in the *Strand* with 'The Japanned Box', 'The Jew's Breast-Plate', 'B. 24', 'The Latin Tutor', and 'The Brown Hand'. (Mar.) Publishes *A Duet with an Occasional Chorus*, a version of his own romance. (Oct.–Dec.) 'The Croxley Master', a boxing story, published in the *Strand*. William Gillette begins 33 years starring in *Sherlock Holmes*, a play by Gillette and ACD.

1900   Accompanies volunteer-staffed Langman hospital as unofficial supervisor to support British forces in the Boer War. (Mar.) Publishes short-story collection, *The Green Flag and other stories of war and sport*. (Oct.) *The Great Boer War* published. Unsuccessful Liberal Unionist parliamentary candidate for Edinburgh Central.

1901   (Aug.) 'The Hound of the Baskervilles' begins serialization in the *Strand*, subtitled 'Another Adventure of Sherlock Holmes'.

1902   (Jan.) *The War in South Africa: Its Cause and Conduct* published. 'Sherlockian' higher criticism begun by Frank Sidgwick in the *Cambridge Review* (23 Jan.). (Mar.) *The Hound of the Baskervilles* published by Newnes. ACD accepts knighthood with reluctance.

1903   (Sept.) *Adventures of Gerard* published by Newnes (previously serialized in the *Strand*). (Oct.) 'The Return of

Sherlock Holmes' begins in the *Strand*. Author's Edition of ACD's major works published in twelve volumes by Smith, Elder and thirteen by D. Appleton & Co. of New York, with prefaces by ACD; many titles omitted.

1904   'Return of Sherlock Holmes' continues in the *Strand*; series designed to conclude with 'The Abbey Grange' (Sept.), but ACD develops earlier allusions and produces 'The Second Stain' (Dec.).

1905   (Mar.) *The Return of Sherlock Holmes* published by Newnes. (Dec.) Serialization of 'Sir Nigel' begun in the *Strand* (concluded Dec. 1906).

1906   (Nov.) Book publication of *Sir Nigel*. ACD defeated as Unionist candidate for Hawick District in general election. (4 July) Death of Louisa ('Touie'), Lady Conan Doyle. ACD deeply affected.

1907   ACD clears the name of George Edalji (convicted in 1903 of cattle-maiming). (18 Sept.) Marries Jean Leckie. (Nov.) Publishes *Through the Magic Door*, a celebration of his literary mentors (earlier version serialized in *Great Thoughts*, 1894).

1908   Moves to Windlesham, Crowborough, Sussex. (Jan.) Death of Sidney Paget. (Sept.) *Round the Fire Stories* published, including some not in earlier *Strand* series. (Sept.– Oct.) 'The Singular Experience of Mr John Scott Eccles' (later retitled as 'The Adventure of Wisteria Lodge') begins occasional series of Holmes stories in the *Strand*.

1909   ACD becomes President of the Divorce Law Reform Union (until 1919). Denis Percy Stewart Conan Doyle born. Takes up agitation against Belgian oppression in the Congo.

1910   (Sept.) 'The Marriage of the Brigadier', the last Gerard story, published in the *Strand*, and (Dec.) the Holmes story of 'The Devil's Foot'. ACD takes six-month lease on Adelphi Theatre; the play *The Speckled Band* opens there, eventually running to 346 performances. Adrian Malcolm Conan Doyle born.

1911   (Apr.) *The Last Galley* (short stories, mostly historical) published. Two more Holmes stories appear in the *Strand*: 'The Red Circle' (Mar., Apr.) and 'The Disappearance of Lady Frances Carfax' (Dec.). ACD declares for Irish Home Rule, under the influence of Sir Roger Casement.

1912 (Apr.–Nov.) The first Professor Challenger story, *The Lost World*, published in the *Strand*, book publication in Oct. Jean Lena Annette Conan Doyle (afterwards Air Commandant Dame Jean Conan Doyle, Lady Bromet) born.

1913 (Feb.) Writes 'Great Britain and the Next War' (*Fortnightly Review*). (Aug.) Second Challenger story, *The Poison Belt*, published. (Dec.) 'The Dying Detective' published in the *Strand*. ACD campaigns for a channel tunnel.

1914 (July) 'Danger!', warning of the dangers of a war-time blockade of Britain, published in the *Strand*. (4 Aug.) Britain declares war on Germany; ACD forms local volunteer force.

1914–15 (Sept.) *The Valley of Fear* begins serialization in the *Strand* (concluding May 1915).

1915 (27 Feb.) *The Valley of Fear* published by George H. Doran in New York. (June) *The Valley of Fear* published in London by Smith, Elder (transferred with rest of ACD stock to John Murray when the firm is sold on the death of Reginald Smith). Five Holmes films released in Germany (ten more during the war).

1916 (Apr., May) First instalments of *The British Campaign in France and Flanders 1914* appear in the *Strand*. (Aug.) *A Visit to Three Fronts* published. Sir Roger Casement convicted of high treason after Dublin Easter Week Rising and executed despite appeals for clemency by ACD and others.

1917 War censor interdicts ACD's history of the 1916 campaigns in the *Strand*. (Sept.) 'His Last Bow' published in the *Strand*. (Oct.) *His Last Bow* published by John Murray (includes 'The Cardboard Box').

1918 (Apr.) ACD publishes *The New Revelation*, proclaiming himself a Spiritualist. (Dec.) *Danger! and other stories* published. Permitted to resume accounts of 1916 and 1917 campaigns in the *Strand*, but that for 1918 never serialized. Death of eldest son, Captain Kingsley Conan Doyle, from influenza aggravated by war wounds.

1919 Death of Brigadier-General Innes Doyle, from post-war pneumonia.

1920–30 ACD engaged in world-wide crusade for Spiritualism.

1921–2 ACD's one-act play, *The Crown Diamond*, tours with Dennis Neilson-Terry as Holmes.

1921 (Oct.) 'The Mazarin Stone' (apparently based on *The Crown Diamond*) published in the *Strand*. Death of mother, Mary Foley Doyle.

1922 (Feb.–Mar.) 'The Problem of Thor Bridge' in the *Strand*. (July) John Murray publishes a collected edition of the non-Holmes short stories in six volumes: *Tales of the Ring and the Camp*, *Tales of Pirates and Blue Water*, *Tales of Terror and Mystery*, *Tales of Twilight and the Unseen*, *Tales of Adventure and Medical Life*, and (Nov.) *Tales of Long Ago* (Sept.) Collected edition of ACD's *Poems* published by Murray.

1923 (Mar.) 'The Creeping Man' published in the *Strand*.

1924 (Jan.) 'The Sussex Vampire' appears in the *Strand*. (June) 'How Watson Learned the Trick', ACD's own Holmes pastiche, appears in *The Book of the Queen's Dolls' House Library*. (Sept.) *Memories and Adventures* published (reprinted with additions and deletions 1930).

1925 (Jan.) 'The Three Garridebs' and (Feb.–Mar.) 'The Illustrious Client' published in the *Strand*. (July) *The Land of Mist*, a Spiritualist novel featuring Challenger, begins serialization in the *Strand*.

1926 (Mar.) *The Land of Mist* published. *Strand* publishes 'The Three Gables' (Oct.), 'The Blanched Soldier' (Nov.), and 'The Lion's Mane' (Dec.).

1927 *Strand* publishes 'The Retired Colourman' (Jan.), 'The Veiled Lodger' (Feb.), and 'Shoscombe Old Place' (Apr.). (June) Murray publishes *The Case-Book of Sherlock Holmes*.

1928 (Oct.) *The Complete Sherlock Holmes Short Stories* published by Murray.

1929 (June) *The Conan Doyle Stories* (containing the six separate volumes issued by Murray in 1922) published. (July) *The Maracot Deep and other stories*, ACD's last collection of his fictional work.

1930 (7 July, 8.30 a.m.) Death of Arthur Conan Doyle. 'Education never ends, Watson. It is a series of lessons with the greatest for the last' ('The Red Circle').

The Sign
of the Four

· **CHAPTER 1** ·

The Science of Deduction

S HERLOCK HOLMES* took his bottle from the corner of the mantelpiece, and his hypodermic syringe* from its neat morocco case. With his long, white, nervous fingers he adjusted the delicate needle, and rolled back his left shirt-cuff. For some little time his eyes rested thoughtfully upon the sinewy forearm and wrist, all dotted and scarred with innumerable puncture-marks. Finally, he thrust the sharp point home, pressed down the tiny piston, and sank back into the velvet-lined arm-chair with a long sigh of satisfaction.

Three times a day for many months I had witnessed this performance, but custom had not reconciled my mind to it. On the contrary, from day to day I had become more irritable at the sight, and my conscience swelled nightly within me at the thought that I had lacked the courage to protest. Again and again I had registered a vow that I should deliver my soul upon the subject; but there was that in the cool, nonchalant air of my companion which made him the last man with whom one would care to take anything approaching to a liberty. His great powers, his masterly manner, and the experience which I had had of his many extraordinary qualities, all made me diffident and backward in crossing him.

Yet upon that afternoon, whether it was the Beaune* which I had taken with my lunch, or the additional exasperation produced by the extreme deliberation of his manner, I suddenly felt that I could hold out no longer.

'Which is it to-day,' I asked, 'morphine or cocaine?'*

He raised his eyes languidly from the old black-letter volume* which he had opened.

'It is cocaine,' he said, 'a seven-per-cent solution. Would you care to try it?'

'No, indeed,' I answered brusquely. 'My constitution has not got over the Afghan campaign* yet. I cannot afford to throw any extra strain upon it.'

He smiled at my vehemence. 'Perhaps you are right, Watson,' he said. 'I suppose that its influence is physically a bad one. I find it, however, so transcendently stimulating and clarifying to the mind that its secondary action is a matter of small moment.'

'But consider!' I said earnestly. 'Count the cost!* Your brain may, as you say, be roused and excited, but it is a pathological and morbid process, which involves increased tissue-change and may at least leave a permanent weakness. You know, too, what a black reaction comes upon you. Surely the game is hardly worth the candle.* Why should you, for a mere passing pleasure, risk the loss of those great powers with which you have been endowed? Remember that I speak not only as one comrade to another, but as a medical man to one for whose constitution he is to some extent answerable.'

He did not seem offended. On the contrary, he put his finger-tips together, and leaned his elbows on the arms of his chair, like one who has a relish for conversation.

'My mind', he said, 'rebels at stagnation. Give me problems, give me work, give me the most abstruse cryptogram or the most intricate analysis, and I am in my own proper atmosphere. I can dispense then with artificial stimulants. But I abhor the dull routine of existence. I crave for mental exaltation. That is why I have chosen my own particular profession, or rather created it, for I am the only one in the world.'

'The only unofficial detective?' I said, raising my eyebrows.

'The only unofficial consulting detective,' he answered. 'I am the last and highest court of appeal in detection. When Gregson or Lestrade or Athelney Jones* are out of their depths—which, by the way, is their normal state—the matter is laid before me. I examine the data, as an expert, and pronounce a specialist's opinion. I claim no credit in

4

such cases. My name figures in no newspaper. The work itself, the pleasure of finding a field for my peculiar powers, is my highest reward. But you have yourself had some experience of my methods of work in the Jefferson Hope case.'

'Yes, indeed,' said I cordially. 'I was never so struck by anything in my life. I even embodied it in a small brochure, with the somewhat fantastic title of "A Study in Scarlet".'*

He shook his head sadly.

'I glanced over it,' said he. 'Honestly, I cannot congratulate you upon it. Detection is, or ought to be, an exact science, and should be treated in the same cold and unemotional manner. You have attempted to tinge it with romanticism, which produces much the same effect as if you worked a love-story or an elopement into the fifth proposition of Euclid.'*

'But the romance was there,' I remonstrated. 'I could not tamper with the facts.'

'Some facts should be suppressed, or, at least, a just sense of proportion should be observed in treating them. The only point in the case which deserved mention was the curious analytical reasoning from effects to causes, by which I succeeded in unravelling it.'

I was annoyed at this criticism of a work which had been specially designed to please him. I confess, too, that I was irritated by the egotism which seemed to demand that every line of my pamphlet should be devoted to his own special doings. More than once during the years that I had lived with him in Baker Street I had observed that a small vanity underlay my companion's quiet and didactic manner. I made no remark, however, but sat nursing my wounded leg. I had had a Jezail bullet* through it some time before, and though it did not prevent me from walking, it ached wearily at every change of the weather.

'My practice has extended recently to the Continent,' said Holmes after a while, filling up his old briar-root pipe. 'I was consulted last week by François le Villard,* who, as you probably know, has come rather to the front lately in the

5

French detective service. He has all the Celtic power of quick intuition, but he is deficient in the wide range of exact knowledge which is essential to the higher developments of his art. The case was concerned with a will and possessed some features of interest. I was able to refer him to two parallel cases, the one at Riga in 1857, and the other at St Louis in 1871, which have suggested to him the true solution. Here is the letter which I had this morning acknowledging my assistance.'

He tossed over, as he spoke, a crumpled sheet of foreign notepaper. I glanced my eyes down it, catching a profusion of notes of admiration, with stray *magnifiques, coup-de-maîtres*, and *tours-de-force*,* all testifying to the ardent admiration of the Frenchman.

'He speaks as a pupil to his master,' said I.

'Oh, he rates my assistance too highly,' said Sherlock Holmes lightly. 'He has considerable gifts himself. He possesses two out of the three qualities necessary for the ideal detective. He has the power of observation and that of deduction. He is only wanting in knowledge, and that may come in time. He is now translating my small works into French.'

'Your works?'

'Oh, didn't you know?' he cried, laughing. 'Yes, I have been guilty of several monographs. They are all upon technical subjects. Here, for example, is one "Upon the Distinction between the Ashes of the Various Tobaccos". In it I enumerate a hundred and forty forms of cigar, cigarette, and pipe tobacco, with coloured plates illustrating the difference in the ash. It is a point which is continually turning up in criminal trials, and which is sometimes of supreme importance as a clue. If you can say definitely, for example, that some murder had been done by a man who was smoking an Indian lunkah,* it obviously narrows your field of search. To the trained eye there is as much difference between the black ash of a Trichinopoly* and the white fluff of bird's-eye* as there is between a cabbage and a potato.'

6

'You have an extraordinary genius for minutiae,' I remarked.

'I appreciate their importance. Here is my monograph upon the tracing of footsteps, with some remarks upon the uses of plaster of Paris as a preserver of impresses. Here, too, is a curious little work upon the influence of a trade upon the form of the hand, with lithotypes of the hands of slaters, sailors, cork-cutters, compositors, weavers, and diamond-polishers. That is a matter of great practical interest to the scientific detective—especially in cases of unclaimed bodies, or in discovering the antecedents of criminals. But I weary you with my hobby.'

'Not at all,' I answered earnestly. 'It is of the greatest interest to me, especially since I have had the opportunity of observing your practical application of it. But you spoke just now of observation and deduction. Surely the one to some extent implies the other.'

'Why, hardly,' he answered, leaning back luxuriously in his armchair and sending up thick blue wreaths from his pipe. 'For example, observation shows me that you have been to the Wigmore Street Post-Office this morning, but deduction lets me know that when there you despatched a telegram.'

'Right!' said I. 'Right on both points! But I confess that I don't see how you arrived at it. It was a sudden impulse upon my part, and I have mentioned it to no one.'

'It is simplicity itself,' he remarked, chuckling at my surprise—'so absurdly simple that an explanation is superfluous; and yet it may serve to define the limits of observation and of deduction. Observation tells me that you have a little reddish mould adhering to your instep. Just opposite the Wigmore Street Office* they have taken up the pavement and thrown up some earth, which lies in such a way that it is difficult to avoid treading in it in entering. The earth is of this peculiar reddish tint which is found, as far as I know, nowhere else in the neighbourhood. So much is observation. The rest is deduction.'

'How, then, did you deduce the telegram?'

'Why, of course I knew that you had not written a letter, since I sat opposite to you all morning. I see also in your open desk there that you have a sheet of stamps and a thick bundle of postcards. What could you go into the post-office for, then, but to send a wire? Eliminate all other factors,* and the one which remains must be the truth.'

'In this case it certainly is so,' I replied, after a little thought. 'The thing, however, is, as you say, of the simplest. Would you think me impertinent if I were to put your theories to a more severe test?'

'On the contrary,' he answered, 'it would prevent me from taking a second dose of cocaine. I should be delighted to look into any problem which you might submit to me.'

'I have heard you say that it is difficult for a man to have any object in daily use without leaving the impress of his individuality upon it in such a way that a trained observer might read it. Now, I have here a watch which has recently come into my possession. Would you have the kindness to let me have an opinion upon the character or habits of the late owner?'

I handed him over the watch with some slight feeling of amusement in my heart, for the test was, as I thought, an impossible one, and I intended it as a lesson against the somewhat dogmatic tone which he occasionally assumed. He balanced the watch in his hand, gazed hard at the dial, opened the back, and examined the works, first with his naked eyes and then with a powerful convex lens. I could hardly keep from smiling at his crestfallen face when he finally snapped the case to and handed it back.

'There are hardly any data,' he remarked. 'The watch has been recently cleaned, which robs me of my most suggestive facts.'

'You are right,' I answered. 'It was cleaned before being sent to me.'

In my heart I accused my companion of putting forward a most lame and impotent excuse to cover his failure. What data could he expect from an uncleaned watch?

8

'Though unsatisfactory, my research has not been entirely barren,' he observed, staring up at the ceiling with dreamy, lack-lustre eyes. 'Subject to your correction, I should judge that the watch belonged to your elder brother, who inherited it from your father.'

'That you gather, no doubt, from the H.W. upon the back?'

'Quite so. The W. suggests your own name. The date of the watch is nearly fifty years back, and the initials are as old as the watch: so it was made for the last generation. Jewellery usually descends to the eldest son, and he is most likely to have the same name as the father. Your father has, if I remember right, been dead many years. It has, therefore, been in the hands of your eldest brother.'

'Right, so far,' said I. 'Anything else?'

'He was a man of untidy habits—very untidy and careless. He was left with good prospects, but he threw away his chances, lived for some time in poverty with occasional short intervals of prosperity, and finally, taking to drink, he died. That is all I can gather.'

I sprang from my chair and limped impatiently about the room with considerable bitterness in my heart.

'This is unworthy of you, Holmes,' I said. 'I could not have believed that you would have descended to this. You have made inquiries into the history of my unhappy brother, and you now pretend to deduce this knowledge in some fanciful way. You cannot expect me to believe that you have read all this from his old watch! It is unkind and, to speak plainly, has a touch of charlatanism in it.'

'My dear doctor,' said he kindly, 'pray accept my apologies. Viewing the matter as an abstract problem, I had forgotten how personal and painful a thing it might be to you. I assure you, however, that I never even knew that you had a brother until you handed me the watch.'

'Then how in the name of all that is wonderful did you get these facts? They are absolutely correct in every particular.'

9

'Ah, that is good luck. I could only say what was the balance of probability. I did not at all expect to be so accurate.'

'But it was not mere guess-work?'

'No, no: I never guess. It is a shocking habit—destructive to the logical faculty. What seems strange to you is only so because you do not follow my train of thought or observe the small facts upon which large inferences may depend. For example, I began by stating that your brother was careless. When you observe the lower part of that watch-case you notice that it is not only dinted in two places, but it is cut and marked all over from the habit of keeping other hard objects, such as coins or keys, in the same pocket. Surely it is no great feat to assume that a man who treats a fifty-guinea* watch so cavalierly must be a careless man. Neither is it a very far-fetched inference that a man who inherits one article of such value is pretty well provided for in other respects.'

I nodded, to show that I followed his reasoning.

'It is very customary for pawnbrokers in England, when they take a watch, to scratch the numbers of the ticket with a pin-point upon the inside of the case. It is more handy than a label, as there is no risk of the number being lost or transposed.' There are no less than four such numbers visible to my lens on the inside of this case. Inference—that your brother was often at low water. Secondary inference—that he had occasional bursts of prosperity, or he could not have redeemed the pledge. Finally, I ask you to look at the inner plate, which contains the keyhole. Look at the thousands of scratches all round the hole—marks where the key has slipped. What sober man's key could have scored those grooves? But you will never see a drunkard's watch without them. He winds it at night, and he leaves these traces of his unsteady hand. Where is the mystery in all this?'

'It is as clear as daylight,' I answered. 'I regret the injustice which I did you. I should have had more faith in your marvellous faculty. May I ask whether you have any professional inquiry on foot at present?'

'None. Hence the cocaine. I cannot live without brain-work. What else is there to live for? Stand at the window here. Was ever such a dreary, dismal, unprofitable world? See how the yellow fog swirls down the street and drifts across the dun-coloured houses. What could be more hopelessly prosaic and material? What is the use of having powers, doctor, when one has no field upon which to exert them? Crime is commonplace, existence is commonplace, and no qualities save those which are commonplace have any function upon earth.'

I had opened my mouth to reply to this tirade, when, with a crisp knock, our landlady entered, bearing a card upon the brass salver.

'A young lady for you, sir,' she said, addressing my companion.

'Miss Mary Morstan,'* he read. 'Hum! I have no recollection of the name. Ask the young lady to step up, Mrs Hudson. Don't go, doctor. I should prefer that you remain.'

· CHAPTER 2 ·

The Statement of the Case

MISS MORSTAN entered the room with a firm step and an outward composure of manner. She was a blonde young lady, small, dainty, well gloved, and dressed in the most perfect taste. There was, however, a plainness and simplicity about her costume which bore with it a suggestion of limited means. The dress was a sombre grayish beige, untrimmed and unbraided, and she wore a small turban of the same dull hue, relieved only by a suspicion of white feather in the side. Her face had neither regularity of feature nor beauty of complexion, but her expression was sweet and amiable, and her large blue eyes were singularly spiritual and sympathetic. In an experience of women which extends over many nations and three separate continents, I have never looked upon a face which gave a clearer promise of a refined and sensitive nature. I could not but observe that as she took the seat which Sherlock Holmes placed for her, her lip trembled, her hand quivered, and she showed every sign of intense inward agitation.

'I have come to you, Mr Holmes,' she said, 'because you once enabled my employer, Mrs Cecil Forrester, to unravel a little domestic complication. She was much impressed by your kindness and skill.'

'Mrs Cecil Forrester,' he repeated thoughtfully. 'I believe that I was of some slight service to her. The case, however, as I remember it, was a very simple one.'

'She did not think so. But at least you cannot say the same of mine. I can hardly imagine anything more strange, more utterly inexplicable, than the situation in which I find myself.'

Holmes rubbed his hands, and his eyes glistened. He leaned forward in his chair with an expression of extraordinary concentration upon his clear-cut, hawk-like features.

'State your case,' said he, in brisk business tones.

I felt that my position was an embarrassing one.

'You will, I am sure, excuse me,' I said, rising from my chair.

To my surprise, the young lady held up her gloved hand to detain me.

'If your friend', she said, 'would be good enough to stop, he might be of inestimable service to me.'

I relapsed into my chair.

'Briefly,' she continued, 'the facts are these. My father was an officer in an Indian regiment, who sent me home when I was quite a child. My mother was dead, and I had no relative in England. I was placed, however, in a comfortable boarding establishment at Edinburgh, and there I remained until I was seventeen years of age. In the year 1878 my father, who was senior captain of his regiment, obtained twelve months' leave and came home. He telegraphed to me from London that he had arrived all safe, and directed me to come down at once, giving the Langham Hotel* as his address. His message, as I remember, was full of kindness and love. On reaching London I drove to the Langham, and was informed that Captain Morstan was staying there, but that he had gone out the night before and had not returned. I waited all day without news of him. That night, on the advice of the manager of the hotel, I communicated with the police, and next morning we advertised in all the papers. Our inquiries led to no result; and from that day to this no word has ever been heard of my unfortunate father. He came home with his heart full of hope to find some peace, some comfort, and instead—'

She put her hand to her throat, and a choking sob cut short the sentence.

'The date?' asked Holmes, opening his note-book.

'He disappeared upon the 3rd of December, 1878—nearly ten years ago.'

'His luggage?'

'Remained at the hotel. There was nothing in it to suggest a clue—some clothes, some books, and a considerable number of curiosities from the Andaman Islands.* He had been one of the officers in charge of the convict-guard there.'

'Had he any friends in town?'

'Only one that we know of—Major Sholto, of his own regiment, the 34th Bombay Infantry. The major had retired some little time before, and lived at Upper Norwood. We communicated with him, of course, but he did not even know that his brother officer was in England.'

'A singular case,' remarked Holmes.

'I have not yet described to you the most singular part. About six years ago—to be exact, upon the 4th of May, 1882—an advertisement appeared in *The Times* asking for the address of Miss Mary Morstan, and stating that it would be to her advantage to come forward. There was no name or address appended. I had at that time just entered the family of Mrs Cecil Forrester in the capacity of governess. By her advice I published my address in the advertisement

13

column. The same day there arrived through the post a small cardboard box addressed to me, which I found to contain a very large lustrous pearl. No word of writing was enclosed. Since then every year upon the same date there has always appeared a similar box, containing a similar pearl, without any clue as to the sender. They have been pronounced by an expert to be of a rare variety and of considerable value. You can see for yourself that they are very handsome.'

She opened a flat box as she spoke and showed me six of the finest pearls that I had ever seen.

'Your statement is most interesting,' said Sherlock Holmes. 'Has anything else occurred to you?'

'Yes, and no later than to-day. That is why I have come to you. This morning I received this letter, which you will perhaps read for yourself.'

'Thank you,' said Holmes. 'The envelope, too, please. Post-mark, London, SW. Date, September 7.* Hum! Man's thumb-mark on corner—probably postman. Best quality paper. Envelopes at sixpence* a packet. Particular man in his stationery. No address. "Be at the third pillar from the left outside the Lyceum Theatre* to-night at seven o'clock. If you are distrustful bring two friends. You are a wronged woman, and shall have justice. Do not bring police. If you do, all will be in vain. Your unknown friend." Well, really, this is a very pretty little mystery! What do you intend to do, Miss Morstan?'

'That is exactly what I want to ask you.'

'Then we shall most certainly go—you and I and—yes, why Dr Watson is the very man. Your correspondent says two friends. He and I have worked together before.'

'But would he come?' she asked with something appealing in her voice and expression.

'I shall be proud and happy,' said I, fervently, 'if I can be of any service.'

'You are both very kind,' she answered. 'I have led a retired life, and have no friends whom I could appeal to. If I am here at six it will do, I suppose?'

'You must not be later,' said Holmes. 'There is one other

14

point, however. Is this handwriting the same as that upon the pearl-box addresses?'

'I have them here,' she answered, producing half a dozen pieces of paper.

'You are certainly a model client. You have the correct intuition. Let us see, now.' He spread out the papers upon the table and gave little darting glances from one to the other. 'They are disguised hands, except the letter,' he said presently; 'but there can be no question as to the authorship. See how the irrepressible Greek *e* will break out, and see the twirl of the final *s*. They are undoubtedly by the same person. I should not like to suggest false hopes, Miss Morstan, but is there any resemblance between this hand and that of your father?'

'Nothing could be more unlike.'

'I expected to hear you say so. We shall look out for you, then, at six. Pray allow me to keep the papers. I may look into the matter before then. It is only half-past three. *Au revoir*, then.'

'*Au revoir*,' said our visitor; and with a bright, kindly glance from one to the other of us, she replaced her pearl-box in her bosom and hurried away.

Standing at the window, I watched her walking briskly down the street until the gray turban and white feather were but a speck in the sombre crowd.

'What a very attractive woman!' I exclaimed, turning to my companion.

He had lit his pipe again, and was leaning back with drooping eyelids. 'Is she?' he said languidly; 'I did not observe.'

'You really are an automaton—a calculating machine,' I cried. 'There is something positively inhuman in you at times.'

He smiled gently.

'It is of the first importance', he said, 'not to allow your judgement to be biased by personal qualities. A client is to me a mere unit, a factor in a problem. The emotional qualities are antagonistic to clear reasoning. I assure you that the most winning woman I ever knew was hanged for poisoning three little children for their insurance-money, and the most repellant man of my acquaintance is a philanthropist who has spent nearly a quarter of a million upon the London poor.'

'In this case, however—'

'I never make exceptions. An exception disproves the rule. Have you ever had occasion to study character in handwriting? What do you make of this fellow's scribble?'*

'It is legible and regular,' I answered. 'A man of business habits and some force of character.'

Holmes shook his head.

'Look at his long letters,' he said. 'They hardly rise above the common herd. That *d* might be an *a*, and that *l* an *e*. Men of character always differentiate their long letters, however illegibly they may write. There is vacillation in his *k*'s and self-esteem in his capitals. I am going out now. I have some few references to make. Let me recommend this book—one of the most remarkable ever penned. It is Winwood Reade's *Martyrdom of Man*.* I shall be back in an hour.'

I sat in the window with the volume in my hand, but my thoughts were far from the daring speculations of the writer. My mind ran upon our late visitor—her smiles, the deep rich tones of her voice, the strange mystery which overhung her life. If she were seventeen at the time of her father's disappearance she must be seven-and-twenty now—a sweet age, when youth has lost its self-consciousness and become a little sobered by experience. So I sat and mused until such dangerous thoughts came into my head that I hurried away to my desk and plunged furiously into the latest treatise upon pathology.* What was I, an army surgeon with a weak leg and a weaker banking account, that I should dare to think of such things? She was a unit, a factor—nothing more. If my future were black, it was better surely to face it like a man than to attempt to brighten it by mere will-o'-the-wisps of the imagination.

· CHAPTER 3 ·

In Quest of a Solution

I T was half-past five before Holmes returned. He was bright, eager, and in excellent spirits, a mood which in his case alternated with fits of the blackest depression.

'There is no great mystery in this matter,' he said, taking the cup of tea which I had poured out for him; 'the facts appear to admit of only one explanation.'

'What! you have solved it already?'

'Well, that would be too much to say. I have discovered a suggestive fact, that is all. It is, however, *very* suggestive. The details are still to be added. I have just found, on consulting the back files of *The Times*,* that Major Sholto, of Upper Norwood, late of the 34th Bombay Infantry, died upon the 28th of April, 1882.'

'I may be very obtuse, Holmes, but I fail to see what this suggests.'

'No? You surprise me. Look at it in this way, then. Captain Morstan disappears. The only person in London whom he could have visited is Major Sholto. Major Sholto denies having heard that he was in London. Four years later Sholto dies. *Within a week of his death* Captain Morstan's daughter receives a valuable present, which is repeated from year to year, and now culminates in a letter which describes her as a wronged woman. What wrong can it refer to except this deprivation of her father? And why should the presents begin immediately after Sholto's death, unless it is that Sholto's heir knows something of the mystery and desires to make compensation? Have you any alternative theory which will meet the facts?'

'But what a strange compensation! And how strangely made! Why, too, should he write a letter now, rather than six years ago? Again, the letter speaks of giving her justice.

What justice can she have? It is too much to suppose that her father is still alive. There is no other injustice in her case that you know of.'

'There are difficulties; there are certainly difficulties,' said Sherlock Holmes pensively; 'but our expedition of to-night will solve them all. Ah, here is a four-wheeler,* and Miss Morstan is inside. Are you all ready? Then we had better go down, for it is a little past the hour.'

I picked up my hat and my heaviest stick, but I observed that Holmes took his revolver from his drawer and slipped it into his pocket. It was clear that he thought that our night's work might be a serious one.

Miss Morstan was muffled in a dark cloak, and her sensitive face was composed, but pale. She must have been more than woman if she did not feel some uneasiness at the strange enterprise upon which we were embarking, yet her self-control was perfect, and she readily answered the few additional questions which Sherlock Holmes put to her.

'Major Sholto was a very particular friend of papa's,' she said. 'His letters were full of allusions to the major. He and papa were in command of the troops at the Andaman Islands, so they were thrown a great deal together. By the way, a curious paper was found in papa's desk which no one could understand. I don't suppose that it is of the slightest importance, but I thought you might care to see it, so I brought it with me. It is here.'

Holmes unfolded the paper carefully and smoothed it out upon his knee. He then very methodically examined it all over with his double lens.

'It is paper of native Indian manufacture,' he remarked. 'It has at some time been pinned to a board. The diagram upon it appears to be a plan of part of a large building with numerous halls, corridors, and passages. At one point is a small cross done in red ink, and above it is "3.37 from left," in faded pencil-writing. In the left-hand corner is a curious hieroglyphic like four crosses in a line with their arms touching. Beside it is written, in very rough and coarse characters, "The sign of the four—Jonathan Small, Mahomet Singh,

Abdullah Khan, Dost Akbar." No, I confess that I do not see how this bears upon the matter. Yet it is evidently a document of importance. It has been kept carefully in a pocket-book; for the one side is as clean as the other.'

'It was in his pocket-book that we found it.'

'Preserve it carefully, then, Miss Morstan, for it may prove to be of use to us. I begin to suspect that this matter may turn out to be much deeper and more subtle than I at first supposed. I must reconsider my ideas.'

He leaned back in the cab, and I could see by his drawn brow and his vacant eye that he was thinking intently. Miss Morstan and I chatted in an undertone about our present expedition and its possible outcome, but our companion maintained his impenetrable reserve until the end of our journey.

It was a September evening* and not yet seven o'clock, but the day had been a dreary one, and a dense drizzly fog lay low upon the great city. Mud-coloured clouds drooped sadly over the muddy streets. Down the Strand the lamps were but misty splotches of diffused light which threw a feeble circular glimmer upon the slimy pavement. The yellow glare from the shop-windows streamed out into the steamy, vaporous air, and threw a murky, shifting radiance across the crowded thoroughfare. There was, to my mind, something eerie and ghost-like in the endless procession of faces which flitted across these narrow bars of light—sad faces and glad, haggard and merry. Like all human kind, they flitted from the gloom into the light, and so back into the gloom once more. I am not subject to impressions, but the dull, heavy evening, with the strange business upon which we were engaged, combined to make me nervous and depressed. I could see from Miss Morstan's manner that she was suffering from the same feeling. Holmes alone could rise superior to petty influences. He held his open notebook upon his knee, and from time to time he jotted down figures and memoranda in the light of his pocket-lantern.

At the Lyceum Theatre the crowds were already thick at the side-entrances. In front a continuous stream of hansoms* and four-wheelers were rattling up, discharging their cargoes of shirt-fronted men and be-shawled, be-diamonded women. We had hardly reached the third pillar, which was our rendezvous, before a small, dark, brisk man in the dress of a coachman accosted us.

'Are you the parties who come with Miss Morstan?' he asked.

'I am Miss Morstan, and these two gentlemen are my friends,' said she.

He bent a pair of wonderfully penetrating and questioning eyes upon us.

'You will excuse me, miss,' he said, with a certain dogged manner, 'but I was to ask you to give me your word that neither of your companions is a police-officer.'

'I give you my word on that,' she answered.

He gave a shrill whistle, on which a street arab* led across a four-wheeler and opened the door. The man who had addressed us mounted to the box, while we took our places inside. We had hardly done so before the driver whipped up his horse, and we plunged away at a furious pace through the foggy streets.

The situation was a curious one. We were driving to an unknown place, on an unknown errand. Yet our invitation was either a complete hoax—which was an inconceivable hypothesis—or else we had good reason to think that important issues might hang upon our journey. Miss Morstan's demeanour was as resolute and collected as ever. I endeavoured to cheer and amuse her by reminiscences of my adventures in Afghanistan; but, to tell the truth, I was myself so excited at our situation, and so curious as to our destination, that my stories were slightly involved. To this day she declares that I told her one moving anecdote as to how a musket looked into my tent at the dead of night, and how I fired a double-barrelled tiger cub at it. At first I had some idea as to the direction in which we were driving; but soon, what with our pace, the fog, and my own limited knowledge of London, I lost my bearings, and knew nothing, save that we seemed to be going a very long way. Sherlock

Holmes was never at fault, however, and he muttered the names as the cab rattled through squares and in and out by tortuous by-streets.

'Rochester Row,' said he. 'Now Vincent Square. Now we come out on the Vauxhall Bridge Road. We are making for the Surrey side, apparently. Yes, I thought so. Now we are on the bridge. You can catch glimpses of the river.'

We did indeed get a fleeting view of a stretch of the Thames, with the lamps shining upon the broad, silent water; but our cab dashed on, and was soon involved in a labyrinth of streets upon the other side.

'Wordsworth Road,' said my companion. 'Priory Road. Lark Hall Lane. Stockwell Place. Robert Street.* Cold Harbour Lane. Our quest does not appear to take us to very fashionable regions.'

We had indeed reached a questionable and forbidding neighbourhood. Long lines of dull brick houses were only relieved by the coarse glare and tawdry brilliancy of public-houses at the corner. Then came rows of two-storeyed villas, each with a fronting of miniature garden, and then again interminable lines of new, staring brick buildings—the monster tentacles which the giant city was throwing out into the country. At last the cab drew up at the third house in a new terrace. None of the other houses were inhabited, and that at which we stopped was as dark as its neighbours, save for a single glimmer in the kitchen-window. On our knocking, however, the door was instantly thrown open by a Hindoo* servant, clad in a yellow turban, white loose- fitting clothes, and a yellow sash. There was something strangely incongruous in this Oriental figure framed in the commonplace doorway of a third-rate suburban dwelling-house.

'The Sahib* awaits you,' said he, and even as he spoke, there came a high, piping voice from some inner room.

'Show them in to me, *khitmutgar*,'* it said. 'Show them straight in to me.'

21

· **CHAPTER 4** ·

The Story of the Bald-Headed Man

W E followed the Indian down a sordid and common passage, ill-lit and worse furnished, until he came to a door upon the right, which he threw open. A blaze of yellow light streamed out upon us, and in the centre of the glare there stood a small man with a very high head, a bristle of red hair all round the fringe of it, and a bald, shining scalp which shot out from among it like a mountain-peak from fir-trees.* He writhed his hands together as he stood, and his features were in a perpetual jerk—now smiling, now scowling, but never for an instant in repose. Nature had given him a pendulous lip, and a too visible line of yellow and irregular teeth,* which he strove feebly to conceal by constantly passing his hand over the lower part of his face. In spite of his obtrusive baldness, he gave the impression of youth. In point of fact, he had just turned his thirtieth year.

'Your servant, Miss Morstan,' he kept repeating, in a thin, high voice. 'Your servant, gentlemen. Pray step into my little sanctum. A small place, miss, but furnished to my own liking. An oasis of art in the howling desert of South London.'

We were all astonished by the appearance of the apartment into which he invited us. In that sorry house it looked as out of place as a diamond of the first water in a setting of brass. The richest and glossiest of curtains and tapestries draped the walls, looped back here and there to expose some richly-mounted painting or Oriental vase. The carpet was of amber and black, so soft and so thick that the foot sank pleasantly into it, as into a bed of moss. Two great tiger-skins thrown athwart it increased the suggestion of Eastern luxury, as did a huge hookah* which stood upon a mat in the corner. A lamp in the fashion of a silver dove was hung from an almost invisible golden wire in the centre of the room. As it burned it filled the air with a subtle and aromatic odour.

'Mr Thaddeus Sholto,' said the little man, still jerking and smiling. 'That is my name. You are Miss Morstan, of course. And these gentlemen—'

'This is Mr Sherlock Holmes, and this Dr Watson.'

'A doctor, eh?' cried he, much excited. 'Have you your stethoscope? Might I ask you—would you have the kindness? I have grave doubts as to my mitral valve,* if you would be so very good. The aortic* I may rely upon, but I should value your opinion upon the mitral.'

I listened to his heart, as requested, but was unable to find anything amiss, save, indeed, that he was in an ecstasy of fear, for he shivered from head to foot.

'It appears to be normal,' I said. 'You have no cause for uneasiness.'

'You will excuse my anxiety, Miss Morstan,' he remarked airily. 'I am a great sufferer, and I have long had suspicions as to that valve. I am delighted to hear that they are unwarranted. Had your father, Miss Morstan, refrained from throwing a strain upon his heart, he might have been alive now.'*

I could have struck the man across the face, so hot was I at this callous and offhand reference to so delicate a matter. Miss Morstan sat down, and her face grew white to the lips.

'I knew in my heart that he was dead,' said she.

'I can give you every information,' said he; 'and, what is more, I can do you justice; and I will, too, whatever Brother Bartholomew may say. I am so glad to have your friends here not only as an escort to you but also as witnesses to what I am about to do and say. The three of us can show a bold front to Brother Bartholomew. But let us have no outsiders—no police or officials. We can settle everything satisfactorily among ourselves without any interference. Nothing would annoy Brother Bartholomew more than any publicity.'

He sat down upon a low settee, and blinked at us enquiringly with his weak, watery blue eyes.

'For my part,' said Holmes, 'whatever you may choose to say will go no further.'

I nodded to show my agreement.

'That is well! That is well!' said he. 'May I offer you a glass of Chianti,* Miss Morstan? Or of Tokay?* I keep no other wines. Shall I open a flask? No? Well, then, I trust that you have no objection to tobacco-smoke, to the balsamic odour of the Eastern tobacco. I am a little nervous, and I find my hookah an invaluable sedative.'

He applied a taper to the great bowl, and, the smoke bubbled merrily through the rose-water. We sat all three in a semicircle, with our heads advanced and our chins upon our hands, while the strange, jerky little fellow, with his high, shining head, puffed uneasily in the centre.

'When I first determined to make this communication to you,' said he, 'I might have given you my address; but I feared that you might disregard my request and bring unpleasant people with you. I took the liberty, therefore, of making an appointment in such a way that my man William might be able to see you first. I have complete confidence in his discretion, and he had orders, if he were dissatisfied, to proceed no further in the matter. You will excuse these precautions, but I am a man of somewhat retiring, and I might even say refined, tastes, and there is nothing more unaesthetic than a policeman. I have a natural shrinking from all forms of rough materialism. I seldom come in contact with the rough crowd. I live, as you see, with some little atmosphere of elegance around me. I may call myself a patron of the arts. It is my weakness. The landscape is a genuine Corot,* and, though a connoisseur might perhaps throw a doubt upon that Salvator Rosa,* there cannot be the least question about the Bouguereau.* I am partial to the modern French school.'*

'You will excuse me, Mr Sholto,' said Miss Morstan, 'but I am here at your request to learn something which you desire to tell me. It is very late, and I should desire the interview to be as short as possible.'

'At the best it must take some time,' he answered; 'for we shall certainly have to go to Norwood and see Brother Bartholomew. We shall all go and try if we can get the better of Brother Bartholomew. He is very angry with me for taking the course which has seemed right to me. I had quite high words with him last night. You cannot imagine what a terrible fellow he is when he is angry.'

'If we are to go to Norwood, it would perhaps be as well to start at once,' I ventured to remark.

He laughed until his ears were quite red.

'That would hardly do,' he cried. 'I don't know what he would say if I brought you in that sudden way. No, I must prepare you by showing you how we all stand to each other. In the first place, I must tell you that there are several points in the story of which I am myself ignorant. I can only lay the facts before you as far as I know them myself.

'My father was, as you may have guessed, Major John Sholto, once of the Indian Army. He retired some eleven years ago and came to live at Pondicherry* Lodge in Upper Norwood. He had prospered in India and brought back with him a considerable sum of money, a large collection of valuable curiosities, and a staff of native servants. With these advantages he bought himself a house, and lived in great luxury. My twin-brother Bartholomew and I were the only children.

'I very well remember the sensation which was caused by the disappearance of Captain Morstan. We read the details in the papers, and knowing that he had been a friend of our father's, we discussed the case freely in his presence. He used to join in our speculations as to what could have happened. Never for an instant did we suspect that he had the whole secret hidden in his own breast, that of all men he alone knew the fate of Arthur Morstan.

'We did know, however, that some mystery, some positive danger, overhung our father. He was very fearful of going out alone, and he always employed two prize-fighters* to act as porters at Pondicherry Lodge. Williams, who drove you to-night, was one of them. He was once light-weight champion of England. Our father would never tell us what it was he feared, but he had a most marked aversion to men with wooden legs. On one occasion he actually fired his revolver at a wooden-legged man, who proved to be a harmless tradesman canvassing for orders. We had to pay a large sum to hush the matter up. My brother and I used to think this a mere whim of my father's; but events have since led us to change our opinion.

'Early in 1882 my father received a letter from India which was a great shock to him. He nearly fainted at the breakfast-table when he opened it, and from that day he sickened to his death. What was in the letter we could never discover, but I could see as he held it that it was short and written in a scrawling hand. He had suffered for years from an enlarged spleen, but he now became rapidly worse, and towards the end of April we were informed that he was beyond all hope, and that he wished to make a last communication to us.

'When we entered his room he was propped up with pillows and breathing heavily. He besought us to lock the door and to come upon either side of the bed. Then, grasping our hands, he made a remarkable statement to us in a voice which was broken as much by emotion as by pain. I shall try and give it to you in his own very words.

' "I have only one thing", he said, "which weighs upon my mind at this supreme moment. It is my treatment of poor Morstan's orphan. The cursed greed which has been my besetting sin through life has withheld from her the treasure, half at least of which should have been hers. And yet I have made no use of it myself, so blind and foolish a thing is avarice. The mere feeling of possession has been so dear to me that I could not bear to share it with another. See that chaplet* tipped with pearls beside the quinine bottle. Even that I could not bear to part with, although I had got it out with the design of sending it to her. You, my sons, will give her a fair share of the Agra treasure. But send her nothing—

not even the chaplet—until I am gone. After all, men have been as bad as this and have recovered.

' "I will tell you how Morstan died," he continued. "He had suffered for years from a weak heart, but he concealed it from everyone. I alone knew it. When in India, he and I, through a remarkable chain of circumstances, came into possession of a considerable treasure. I brought it over to England, and on the night of Morstan's arrival he came straight over here to claim his share. He walked over from the station, and was admitted by my faithful old Lal Chowdar, who is now dead. Morstan and I had a difference of opinion as to the division of the treasure, and we came to heated words. Morstan had sprung out of his chair in a paroxysm of anger, when he suddenly pressed his hand to his side, his face turned a dusky hue, and he fell backwards, cutting his head against the corner of the treasure-chest. When I stooped over him I found, to my horror, that he was dead.

' "For a long time I sat half distracted, wondering what I should do. My first impulse was, of course, to call for assistance; but I could not but recognize that there was every chance that I would be accused of his murder. His death at the moment of a quarrel, and the gash in his head, would be black against me. Again, an official inquiry could not be made without bringing out some facts about the treasure, which I was particularly anxious to keep secret. He had told me that no soul upon earth knew where he had gone. There seemed to be no necessity why any soul ever should know.

' "I was still pondering over the matter, when, looking up, I saw my servant, Lal Chowdar, in the doorway. He stole in and bolted the door behind him. 'Do not fear, Sahib,' he said; 'no one need know that you have killed him. Let us hide him away, and who is the wiser?' 'I did not kill him,' said I. Lal Chowdar shook his head and smiled. 'I heard it all, Sahib,' said he; 'I heard you quarrel, and I heard the blow. But my lips are sealed. All are asleep in the house. Let us put him away together.' That was enough to decide me.

If my own servant could not believe my innocence, how could I hope to make it good before twelve foolish trades-men in a jury-box? Lal Chowdar and I disposed of the body that night, and within a few days the London papers were full of the mysterious disappearance of Captain Morstan. You will see from what I say that I can hardly be blamed in the matter. My fault lies in the fact that we concealed not only the body, but also the treasure, and that I have clung to Morstan's share as well as to my own. I wish you, therefore, to make restitution. Put your ears down to my mouth. The treasure is hidden in—"

'At this instant a horrible change came over his expres-sion; his eyes stared wildly, his jaw dropped, and he yelled, in a voice which I can never forget, "Keep him out! For Christ's sake keep him out!" We both stared round at the window behind us upon which his gaze was fixed. A face was looking in at us out of the darkness. We could see the whitening of the nose where it was pressed against the glass. It was a bearded, hairy face, with wild cruel eyes and an expression of concentrated malevolence. My brother and I rushed towards the window, but the man was gone. When we returned to my father his head had dropped and his pulse had ceased to beat.

'We searched the garden that night, but found no sign of the intruder save that just under the window a single footmark was visible in the flower-bed. But for that one trace, we might have thought that our imaginations had conjured up that wild, fierce face. We soon, however, had another and a more striking proof that there were secret agencies at work all round us. The window of my father's room was found open in the morning, his cupboards and boxes had been rifled, and upon his chest was fixed a torn piece of paper, with the words "The sign of the four" scrawled across it. What the phrase meant or who our secret visitor may have been, we never knew. As far as we can judge, none of my father's property had been actually stolen, though everything had been turned out. My brother and I naturally associated this peculiar incident with the fear which haunted my father during his life; but it is still a complete mystery to us.'

The little man stopped to relight his hookah and puffed thoughtfully for a few moments. We had all sat absorbed, listening to his extraordinary narrative. At the short account of her father's death Miss Morstan had turned deadly white, and for a moment I feared that she was about to faint. She rallied, however, on drinking a glass of water which I quietly poured out for her from a Venetian carafe upon the side-table. Sherlock Holmes leaned back in his chair with an abstracted expression and the lids drawn low over his glittering eyes. As I glanced at him I could not but think how on that very day he had complained bitterly of the commonplaceness of life. Here at least was a problem which would tax his sagacity to the utmost. Mr Thaddeus Sholto looked from one to the other of us with an obvious pride at the effect which his story had produced, and then continued between the puffs of his overgrown pipe.

'My brother and I', said he, 'were, as you may imagine, much excited as to the treasure which my father had spoken of. For weeks and for months we dug and delved in every part of the garden without discovering its whereabouts. It was maddening to think that the hiding-place was on his very lips at the moment that he died. We could judge the splendour of the missing riches by the chaplet which he had taken out. Over this chaplet my brother Bartholomew and I had some little discussion. The pearls were evidently of great value, and he was averse to part with them, for, between friends, my brother was himself a little inclined to my father's fault. He thought, too, that if we parted with the chaplet it might give rise to gossip and finally bring us into trouble. It was all that I could do to persuade him to let me find out Miss Morstan's address and send her a detached pearl at fixed intervals, so that at least she might never feel destitute.'

'It was a kindly thought,' said our companion earnestly; 'it was extremely good of you.'

The little man waved his hand deprecatingly.

29

'We were your trustees,' he said; 'that was the view which I took of it, though Brother Bartholomew could not altogether see it in that light. We had plenty of money ourselves. I desired no more. Besides, it would have been such bad taste to have treated a young lady in so scurvy a fashion. "Le mauvais goût mène au crime."* The French have a very neat way of putting these things. Our difference of opinion on this subject went so far that I thought it best to set up rooms for myself; so I left Pondicherry Lodge, taking the old *khitmutgar* and Williams with me. Yesterday, however, I learn that an event of extreme importance has occurred. The treasure has been discovered. I instantly communicated with Miss Morstan, and it only remains for us to drive out to Norwood and demand our share. I explained my views last night to Brother Bartholomew, so we shall be expected, if not welcome, visitors.'

Mr Thaddeus Sholto ceased, and sat twitching on his luxurious settee. We all remained silent, with our thoughts upon the new development which the mysterious business had taken. Holmes was the first to spring to his feet.

'You have done well, sir, from first to last,' said he. 'It is possible that we may be able to make you some small return by throwing some light upon that which is still dark to you. But, as Miss Morstan remarked just now, it is late, and we had best put the matter through without delay.'

Our new acquaintance very deliberately coiled up the tube of his hookah and produced from behind a curtain a very long befrogged topcoat with Astrakhan* collar and cuffs. This he buttoned tightly up in spite of the extreme closeness of the night and finished his attire by putting on a rabbit-skin cap with hanging lappets which covered the ears, so that no part of him was visible save his mobile and peaky face.

'My health is somewhat fragile,' he remarked as he led the way down the passage. 'I am compelled to be a valetudinarian.'*

Our cab was awaiting us outside, and our programme was evidently prearranged, for the driver started off at once at a rapid pace. Thaddeus Sholto talked incessantly in a voice which rose high above the rattle of the wheels.

'Bartholomew is a clever fellow,' said he. 'How do you think he found out where the treasure was? He had come to the conclusion that it was somewhere indoors: so he worked out all the cubic space of the house,* and made measurements everywhere so that not one inch should be unaccounted for. Among other things, he found that the height of the building was seventy-four feet, but on adding together the heights of all the separate rooms and making every allowance for the space between, which he ascertained by borings, he could not bring the total to more than seventy feet. There were four feet unaccounted for. These could only be at the top of the building. He knocked a hole, therefore, in the lath and plaster ceiling of the highest room, and there, sure enough, he came upon another little garret above it, which had been sealed up and was known to no one. In the centre stood the treasure-chest, resting upon two rafters. He lowered it through the hole, and there it lies. He computes the value of the jewels at not less than half a million sterling.'

At the mention of this gigantic sum we all stared at one another open-eyed. Miss Morstan, could we secure her rights, would change from a needy governess to the richest heiress in England. Surely it was the place of a loyal friend to rejoice at such news; yet I am ashamed to say that selfishness took me by the soul, and that my heart turned as heavy as lead within me. I stammered out some few halting words of congratulation, and then sat downcast, with my head drooped, deaf to the babble of our new acquaintance. He was clearly a confirmed hypochondriac, and I was dreamily conscious that he was pouring forth interminable trains of symptoms, and imploring information as to the composition and action of innumerable quack nostrums, some of which he bore about in a leather case in his pocket. I trust that he may not remember any of the answers which I gave him that night. Holmes declares that he overheard me caution him against the great danger of taking more than two drops of castor-oil, while I recommended strych-nine in large doses as a sedative. However that may be, I was certainly relieved when our cab pulled up with a jerk and the coachman sprang down to open the door.

'This, Miss Morstan, is Pondicherry Lodge,' said Mr Thaddeus Sholto, as he handed her out.

· CHAPTER 5 ·

The Tragedy of Pondicherry Lodge

IT was nearly eleven o'clock when we reached this final stage of our night's adventures. We had left the damp fog of the great city behind us, and the night was fairly fine. A warm wind blew from the westward, and heavy clouds moved slowly across the sky, with half a moon peeping occasionally through the rifts. It was clear enough to see for some distance, but Thaddeus Sholto took down one of the side-lamps from the carriage to give us a better light upon our way.

Pondicherry Lodge stood in its own grounds, and was girt round with a very high stone wall topped with broken glass. A single narrow iron-clamped door formed the only means of entrance. On this our guide knocked with a peculiar postman-like rat-tat.

'Who is there?' cried a gruff voice from within.

'It is I, McMurdo. You surely know my knock by this time.'

There was a grumbling sound and a clanking and jarring of keys. The door swung heavily back, and a short, deep-chested man stood in the opening, with the yellow light of the lantern shining upon his protruded face and twinkling, distrustful eyes.

'That you, Mr Thaddeus? But who are the others? I had no orders about them from the master.'

'No, McMurdo? You surprise me! I told my brother last
night that I should bring some friends.'

'He hain't been out o' his room to-day, Mr Thaddeus, and
I have no orders. You know very well that I must stick to
regulations. I can let you in, but your friends they must just
stop where they are.'

This was an unexpected obstacle. Thaddeus Sholto looked
about him in a perplexed and helpless manner.

'This is too bad of you, McMurdo!' he said. 'If I guarantee
them, that is enough for you. There is the young lady, too.
She cannot wait on the public road at this hour.'

'Very sorry, Mr Thaddeus,' said the porter inexorably.
'Folk may be friends o' yours, and yet no friends o' the
master's. He pays me well to do my duty, and my duty I'll
do. I don't know none o' your friends.'

'Oh, yes, you do, McMurdo,' cried Sherlock Holmes
genially. 'I don't think you can have forgotten me. Don't
you remember that amateur* who fought three rounds with
you at Alison's rooms on the night of your benefit* four
years back?'

'Not Mr Sherlock Holmes!' roared the prize-fighter.
'God's truth! how could I have mistook you? If instead o'
standin' there so quiet you had just stepped up and given
me that cross-hit of yours under the jaw, I'd ha' known you
without a question. Ah, you're one that has wasted your
gifts, you have! You might have aimed high, if you had
joined the fancy.'*

'You see, Watson, if all else fails me, I have still one of the
scientific professions open to me,' said Holmes, laughing.
'Our friend won't keep us out in the cold now, I am sure.'

'In you come, sir, in you come—you and your friends,' he
answered. 'Very sorry, Mr Thaddeus, but orders are very
strict. Had to be certain of your friends before I let them in.'

Inside, a gravel path wound through desolate grounds to
a huge clump of a house, square and prosaic, all plunged in
shadow save where a moonbeam struck one corner and
glimmered in a garret window. The vast size of the building,
with its gloom and its deathly silence, struck a chill to the

heart. Even Thaddeus Sholto seemed ill at ease, and the lantern quivered and rattled in his hand.

'I cannot understand it,' he said. 'There must be some mistake. I distinctly told Bartholomew that we should be here, and yet there is no light in his window. I do not know what to make of it.'

'Does he always guard the premises in this way?' asked Holmes.

'Yes; he has followed my father's custom. He was the favourite son, you know, and I sometimes think that my father may have told him more than he ever told me. That is Bartholomew's window up there where the moonshine strikes. It is quite bright, but there is no light from within, I think.'

'None,' said Holmes. 'But I see the glint of a light in that little window beside the door.'

'Ah, that is the housekeeper's room. That is where old Mrs Bernstone sits. She can tell us all about it. But perhaps you would not mind waiting here for a minute or two, for if we all go in together, and she has had no word of our coming, she may be alarmed. But, hush! what is that?'

He held up the lantern, and his hand shook until the circles of light flickered and wavered all round us. Miss Morstan seized my wrist, and we all stood, with thumping hearts, straining our ears. From the great black house there sounded through the silent night the saddest and most pitiful of sounds—the shrill, broken whimpering of a frightened woman.

'It is Mrs Bernstone,' said Sholto. 'She is the only woman in the house. Wait here. I shall be back in a moment.'

He hurried for the door, and knocked in his peculiar way. We could see a tall old woman admit him, and sway with pleasure at the very sight of him.

'Oh, Mr Thaddeus, sir, I am so glad you have come! I am so glad you have come, Mr Thaddeus, sir!'

We heard her reiterated rejoicings until the door was closed and her voice died away into a muffled monotone.

Our guide had left us the lantern. Holmes swung it slowly round and peered keenly at the house, and at the great rubbish-heaps which cumbered the grounds. Miss Morstan and I stood together, and her hand was in mine. A wondrous subtle thing is love, for here were we two, who had never seen each other before that day, between whom no word or even look of affection had ever passed, and yet now in an hour of trouble our hands instinctively sought for each other. I have marvelled at it since, but at the time it seemed the most natural thing that I should go out to her so, and, as she has often told me, there was in her also the instinct to turn to me for comfort and protection. So we stood hand-in-hand, like two children, and there was peace in our hearts for all the dark things that surrounded us.

'What a strange place!' she said, looking round.

'It looks as though all the moles in England had been let loose in it. I have seen something of the sort on the side of a hill near Ballarat,* where the prospectors had been at work.'

'And from the same cause,' said Holmes. 'These are the traces of the treasure-seekers. You must remember that they were six years looking for it. No wonder that the grounds look like a gravel-pit.'

At that moment the door of the house burst open, and Thaddeus Sholto came running out, with his hands thrown forward and terror in his eyes.

'There is something amiss with Bartholomew!' he cried. 'I am frightened! My nerves cannot stand it.'

He was, indeed, half blubbering with fear, and his twitching, feeble face peeping out from the great Astrakhan collar had the helpless, appealing expression of a terrified child.

'Come into the house,' said Holmes in his crisp, firm way.

'Yes, do!' pleaded Thaddeus Sholto. 'I really do not feel equal to giving directions.'

We all followed him into the housekeeper's room, which stood upon the left-hand side of the passage. The old woman was pacing up and down with a scared look and restless, picking fingers, but the sight of Miss Morstan appeared to have a soothing effect upon her.

'God bless your sweet, calm face!' she cried, with an hysterical sob. 'It does me good to see you. Oh, but I have been sorely tried this day!'

Our companion patted her thin, work-worn hand, and murmured some few words of kindly, womanly comfort which brought the colour back into the other's bloodless cheeks.

'Master has locked himself in, and will not answer me,' she explained. 'All day I have waited to hear from him, for he often likes to be alone; but an hour ago I feared that something was amiss, so I went up and peeped through the keyhole. You must go up, Mr Thaddeus—you must go up and look for yourself. I have seen Mr Bartholomew Sholto in joy and in sorrow for ten long years, but I never saw him with such a face on him as that.'

Sherlock Holmes took the lamp and led the way, for Thaddeus Sholto's teeth were chattering in his head. So shaken was he that I had to pass my hand under his arm as we went up the stairs, for his knees were trembling under him. Twice as we ascended, Holmes whipped his lens out of his pocket and carefully examined marks which appeared to me to be mere shapeless smudges of dust upon the coconut matting which served as a stair-carpet. He walked slowly from step to step, holding the lamp low, and shooting keen glances to right and left. Miss Morstan had remained behind with the frightened housekeeper.

The third flight of stairs ended in a straight passage of some length, with a great picture in Indian tapestry upon the right of it and three doors upon the left. Holmes advanced along it in the same slow and methodical way, while we kept close at his heels, with our long black shadows streaming backwards down the corridor. The third door was that which we were seeking. Holmes knocked without receiving any answer, and then tried to turn the handle and force it open. It was locked on the inside, however, and by a broad and powerful bolt, as we could see when we set our lamp up against it. The key being turned, however, the hole was not entirely closed. Sherlock Holmes bent down to it, and instantly rose again with a sharp intaking of the breath.

'There is something devilish in this, Watson,' said he, more moved than I had ever before seen him. 'What do you make of it?'

I stooped to the hole, and recoiled in horror. Moonlight was streaming into the room, and it was bright with a vague and shifty radiance. Looking straight at me, and suspended, as it were, in the air, for all beneath was in shadow, there hung a face—the very face of our companion Thaddeus. There was the same high, shining head, the same circular bristle of red hair, the same bloodless countenance. The features were set, however, in a horrible smile, a fixed and unnatural grin, which in that still and moonlit room was more jarring to the nerves than any scowl or contortion. So like was the face to that of our little friend that I looked round at him to make sure that he was indeed with us. Then I recalled to mind that he had mentioned to us that his brother and he were twins.

'This is terrible!' I said to Holmes. 'What is to be done?'

'The door must come down,' he answered, and, springing against it, he put all his weight upon the lock.

It creaked and groaned, but did not yield. Together we flung ourselves upon it once more, and this time it gave way with a sudden snap, and we found ourselves within Bartholomew Sholto's chamber.

It appeared to have been fitted up as a chemical laboratory. A double line of glass-stoppered bottles was drawn up upon the wall opposite the door, and the table was littered over with Bunsen burners, test-tubes, and retorts. In the corners stood carboys of acid in wicker baskets. One of these appeared to leak or to have been broken, for a stream of dark-coloured liquid had trickled out from it, and the air was heavy with a peculiarly pungent, tar-like odour. A set of steps stood at one side of the room, in the midst of a litter of lath and plaster, and above them there was an opening in the ceiling large enough for a man to pass through. At the foot of the steps a long coil of rope was thrown carelessly together.

By the table, in a wooden armchair, the master of the house was seated all in a heap, with his head sunk upon his left shoulder, and that ghastly, inscrutable smile upon his face. He was stiff and cold, and had clearly been dead many hours. It seemed to me that not only his features, but all his limbs, were twisted and turned in the most fantastic fashion. By his hand upon the table there lay a peculiar instrument—a brown, close-grained stick, with a stone head like a hammer, rudely lashed on with coarse twine. Beside it was a torn sheet of note-paper with some words scrawled upon it. Holmes glanced at it, and then handed it to me.

'You see,' he said, with a significant raising of the eyebrows.

In the light of the lantern I read, with a thrill of horror, 'The sign of the four.'

'In God's name, what does it all mean?' I asked.

'It means murder,' said he, stooping over the dead man. 'Ah! I expected it. Look here!'

He pointed to what looked like a long dark thorn stuck in the skin just above the ear.

'It looks like a thorn,' said I.

'It is a thorn. You may pick it out. But be careful, for it is poisoned.'

I took it up between my finger and thumb. It came away from the skin so readily that hardly any mark was left behind. One tiny speck of blood showed where the puncture had been.

'This is all an insoluble mystery to me,' said I. 'It grows darker instead of clearer.'

'On the contrary,' he answered, 'it clears every instant. I only require a few missing links to have an entirely connected case.'

We had almost forgotten our companion's presence since we entered the chamber. He was still standing in the doorway, the very picture of terror, wringing his hands and moaning to himself. Suddenly, however, he broke out into a sharp, querulous cry.

'The treasure is gone!' he said. 'They have robbed him of the treasure! There is the hole through which we lowered it. I helped him to do it! I was the last person who saw him! I left him here last night, and I heard him lock the door as I came downstairs.'

'What time was that?'

'It was ten o'clock. And now he is dead, and the police will be called in, and I shall be suspected of having had a hand in it. Oh yes, I am sure I shall. But you don't think so, gentlemen? Surely you don't think that it was I? Is it likely that I would have brought you here if it were I? Oh dear! oh dear! I know that I shall go mad!'

He jerked his arms and stamped his feet in a kind of convulsive frenzy.

'You have no reason for fear, Mr Sholto,' said Holmes kindly, putting his hand upon his shoulder; 'take my advice, and drive down to the station to report the matter to the police. Offer to assist them in every way. We shall wait here until your return.'

The little man obeyed in a half-stupefied fashion, and we heard him stumbling down the stairs in the dark.

· CHAPTER 6 ·

Sherlock Holmes Gives a Demonstration

'Now, Watson,' said Holmes, rubbing his hands, 'we have half an hour to ourselves. Let us make good use of it. My case is, as I have told you, almost complete; but we must not err on the side of over-confidence. Simple as the case seems now, there may be something deeper underlying it.'

'Simple!' I ejaculated.

'Surely,' said he, with something of the air of a clinical professor expounding to his class. 'Just sit in the corner there, that your footprints may not complicate matters. Now to work! In the first place, how did these folk come, and how did they go? The door has not been opened since last night.

How of the window?' He carried the lamp across to it, muttering his observations aloud the while but addressing them to himself rather than to me. 'Window is snibbed* on the inner side. Frame-work is solid. No hinges at the side. Let us open it. No water-pipe near. Roof quite out of reach. Yet a man has mounted by the window. It rained a little last night. Here is the print of a foot in mould upon the sill. And here is a circular muddy mark, and here again upon the floor, and here again by the table. See here, Watson! This is really a very pretty demonstration.'

I looked at the round, well-defined muddy discs.

'That is not a foot-mark,' said I.

'It is something much more valuable to us. It is the impression of a wooden stump. You see here on the sill is the boot-mark, a heavy boot with a broad metal heel, and beside it is the mark of the timber-toe.'

'It is the wooden-legged man.'

'Quite so. But there has been someone else—a very able and efficient ally. Could you scale that wall, doctor?'

I looked out of the open window. The moon still shone brightly on that angle of the house. We were a good sixty feet from the ground, and, look where I could, I could see no foothold, nor as much as a crevice in the brickwork.

'It is absolutely impossible,' I answered.

'Without aid it is so. But suppose you had a friend up here who lowered you this good stout rope which I see in the corner, securing one end of it to this great hook in the wall. Then, I think, if you were an active man, you might swarm up, wooden leg and all. You would depart, of course, in the same fashion, and your ally would draw up the rope, untie it from the hook, shut the window, snib it on the inside, and get away in the way that he originally came. As a minor point, it may be noted,' he continued, fingering the rope, 'that our wooden-legged friend, though a fair climber, was not a professional sailor. His hands were far from horny. My lens discloses more than one blood-mark, especially towards the end of the rope, from which I gather that he slipped down with such velocity that he took the skin off his hand.'

'This is all very well,' said I; 'but the thing becomes more unintelligible than ever. How about this mysterious ally? How came he into the room?'

'Yes, the ally!' repeated Holmes pensively. 'There are features of interest about this ally. He lifts the case from the regions of the commonplace. I fancy that this ally breaks fresh ground in the annals of crime in this country—though parallel cases suggest themselves from India and, if my memory serves me, from Senegambia.'*

'How came he, then?' I reiterated. 'The door is locked; the window is inaccessible. Was it through the chimney?'

'The grate is much too small,' he answered. 'I had already considered that possibility.'

'How, then?' I persisted.

'You will not apply my precept,' he said, shaking his head. 'How often have I said to you that when you have eliminated the impossible, whatever remains, *however improbable*, must be the truth?* We know that he did not come through the door, the window, or the chimney. We also know that he could not have been concealed in the room, as there is no concealment possible. Whence, then, did he come?'

'He came through the hole in the roof!' I cried.

'Of course he did. He must have done so. If you will have the kindness to hold the lamp for me, we shall now extend our researches to the room above—the secret room in which the treasure was found.'

He mounted the steps, and, seizing a rafter with either hand, he swung himself up into the garret. Then, lying on his face, he reached down for the lamp, and held it while I followed him.

The chamber in which we found ourselves was about ten feet one way and six the other. The floor was formed by the rafters, with thin lath-and-plaster between, so that in walking one had to step from beam to beam. The roof ran up to an apex, and was evidently the inner shell of the true roof of the house. There was no furniture of any sort, and the accumulated dust of years lay thick upon the floor.

'Here you are, you see,' said Sherlock Holmes, putting his hand against the sloping wall. 'This is a trapdoor which leads out on to the roof. I can press it back, and here is the roof itself, sloping at a gentle angle. This, then, is the way by which Number One entered. Let us see if we can find some other traces of his individuality?'

He held down the lamp to the floor, and as he did so I saw for the second time that night a startled, surprised look come over his face. For myself, as I followed his gaze, my skin was cold under my clothes. The floor was covered thickly with the prints of a naked foot—clear, well-defined, perfectly formed, but scarce half the size of those of an ordinary man.

'Holmes,' I said, in a whisper, 'a child has done this horrid thing.'

He had recovered his self-possession in an instant.

'I was staggered for the moment,' he said, 'but the thing is quite natural. My memory failed me, or I should have been able to foretell it. There is nothing more to be learned here. Let us go down.'

'What is your theory, then, as to those foot-marks?' I asked eagerly, when we had regained the lower room once more.

'My dear Watson, try a little analysis yourself,' said he, with a touch of impatience. 'You know my methods. Apply them, and it will be instructive to compare results.'

'I cannot conceive anything which will cover the facts,' I answered.

'It will be clear enough to you soon,' he said, in an off-hand way. 'I think that there is nothing else of importance here, but I will look.'

He whipped out his lens and a tape measure, and hurried about the room on his knees, measuring, comparing, examining, with his long thin nose only a few inches from the planks and his beady eyes gleaming and deep-set like those of a bird. So swift, silent, and furtive were his movements, like those of a trained bloodhound picking out a scent, that I could not but think what a terrible criminal he would have made had he turned his energy and sagacity against the law, instead of exerting them in its defence. As he hunted about, he kept muttering to himself, and finally he broke out into a loud crow of delight.

'We are certainly in luck,' said he. 'We ought to have very little trouble now. Number One has had the misfortune to tread in the creosote. You can see the outline of the edge of his small foot here at the side of this evil-smelling mess. The carboy has been cracked, you see, and the stuff has leaked out.'

'What then?' I asked.

'Why, we have got him, that's all,' said he. 'I know a dog that would follow that scent to the world's end. If a pack can track a trailed herring across a shire,* how far can a specially trained hound follow so pungent a smell as this? It sounds like a sum in the rule of three.* The answer should give us the—But hallo! here are the accredited representatives of the law.'

Heavy steps and the clamour of loud voices were audible from below, and the hall door shut with a loud crash.

'Before they come,' said Holmes, 'just put your hand here on this poor fellow's arm, and here on his leg. What do you feel?'

'The muscles are as hard as a board,' I answered.

'Quite so. They are in a state of extreme contraction, far exceeding the usual *rigor mortis*.* Coupled with this distortion of the face, this Hippocratic smile, or *"risus sardonicus"*,* as the old writers called it, what conclusion would it suggest to your mind?'

'Death from some powerful vegetable alkaloid,' I answered, 'some strychnine-like substance which would produce tetanus.'*

'That was the idea which occurred to me the instant I saw the drawn muscles of the face. On getting into the room I at once looked for the means by which the poison had entered the system. As you saw, I discovered a thorn which had been driven or shot with no great force into the scalp. You observe that the part struck was that which would be turned towards the hole in the ceiling if the man were erect in his chair. Now examine this thorn.'

I took it up gingerly and held it in the light of the lantern. It was long, sharp, and black, with a glazed look near the point as though some gummy substance had dried upon it. The blunt end had been trimmed and rounded off with a knife.

'Is that an English thorn?' he asked.

'No, it certainly is not.'

'With all these data you should be able to draw some just inference. But here are the regulars: so the auxiliary forces may beat a retreat.'

As he spoke, the steps which had been coming nearer sounded loudly on the passage, and a very stout, portly man in a grey suit strode heavily into the room. He was red-faced, burly, and plethoric, with a pair of very small twinkling eyes which looked keenly out from between swollen and puffy pouches. He was closely followed by an inspector in uniform and by the still palpitating Thaddeus Sholto.

'Here's a business!' he cried in a muffled, husky voice. 'Here's a pretty business! But who are all these? Why, the house seems to be as full as a rabbit-warren!'

'I think you must recollect me, Mr Athelney Jones,' said Holmes quietly.

'Why, of course I do!' he wheezed. 'It's Mr Sherlock Holmes, the theorist. Remember you! I'll never forget how you lectured us all on causes and inferences and effects in the Bishopgate jewel case. It's true you set us on the right track; but you'll own now that it was more by good luck than good guidance.'

'It was a piece of very simple reasoning.'

'Oh, come, now, come! Never be ashamed to own up. But what is all this? Bad business! Bad business! Stern facts here—no room for theories. How lucky that I happened to be out at Norwood over another case! I was at the station when the message arrived. What d'you think the man died of?'

'Oh, this is hardly a case for me to theorize over,' said Holmes dryly.

'No, no. Still, we can't deny that you hit the nail on the head sometimes. Dear me! Door locked, I understand. Jewels worth half a million missing. How was the window?'

'Fastened; but there are steps on the sill.'

'Well, well, if it was fastened the steps could have nothing to do with the matter. That's common-sense. Man might have died in a fit; but then the jewels are missing. Ha! I have a theory. These flashes come upon me at times.—Just step outside, Sergeant, and you, Mr Sholto. Your friend can remain.—What do you think of this, Holmes? Sholto was, on his own confession, with his brother last night. The brother died in a fit, on which Sholto walked off with the treasure? How's that?'

'On which the dead man very considerately got up and locked the door on the inside.'

'Hum! There's a flaw there. Let us apply common-sense to the matter. This Thaddeus Sholto *was* with his brother; there *was* a quarrel: so much we know. The brother is dead and the jewels are gone. So much also we know. No one saw the brother from the time Thaddeus left him. His bed had not been slept in. Thaddeus is evidently in a most disturbed state of mind. His appearance is—well, not attractive. You see that I am weaving my web round Thaddeus. The net begins to close upon him.'

'You are not quite in possession of the facts yet,' said Holmes. 'This splinter of wood, which I have every reason to believe to be poisoned, was in the man's scalp where you still see the mark; this card, inscribed as you see it, was on the table, and beside it lay this rather curious stone-headed instrument. How does all that fit into your theory?'

'Confirms it in every respect,' said the fat detective, pompously. 'House is full of Indian curiosities. Thaddeus brought this up, and if this splinter be poisonous, Thaddeus may as well have made murderous use of it as any other man. The card is some hocus-pocus—a blind, as like as not. The

only question is, how did he depart? Ah, of course, here is
a hole in the roof.'

With great activity, considering his bulk, he sprang up the
steps and squeezed through into the garret, and immediately
afterwards we heard his exulting voice proclaiming that he
had found the trapdoor.

'He can find something,' remarked Holmes, shrugging his
shoulders; 'he has occasional glimmerings of reason. *Il n'y a
pas des sots si incommodes que ceux qui ont de l'esprit!'**

'You see!' said Athelney Jones, reappearing down the steps
again; 'facts are better than theories, after all. My view of
the case is confirmed. There is a trapdoor communicating
with the roof, and it is partly open.'

'It was I who opened it.'

'Oh, indeed! You did notice it, then?' He seemed a little
crestfallen at the discovery. 'Well, whoever noticed it, it
shows how our gentleman got away. Inspector!'

'Yes, sir,' from the passage.

'Ask Mr Sholto to step this way.—Mr Sholto, it is my duty
to inform you that anything which you may say will be used
against you.* I arrest you in the Queen's name as being
concerned in the death of your brother.'

'There, now! Didn't I tell you!' cried the poor little man,
throwing out his hands, and looking from one to the other
of us.

'Don't trouble yourself about it, Mr Sholto,' said Holmes;
'I think that I can engage to clear you of the charge.'

'Don't promise too much, Mr Theorist, don't promise too
much!' snapped the detective. 'You may find it a harder
matter than you think.'

'Not only will I clear him, Mr Jones, but I will make you
a free present of the name and description of one of the two
people who were in this room last night. His name, I have
every reason to believe, is Jonathan Small. He is a poorly-
educated man, small, active, with his right leg off, and
wearing a wooden stump which is worn away upon the
inner side. His left boot has a coarse, square-toed sole, with
an iron band round the heel. He is a middle-aged man,

much sunburned, and has been a convict. These few indications may be of some assistance to you, coupled with the fact that there is a good deal of skin missing from the palm of his hand. The other man—'

'Ah! the other man?' asked Athelney Jones in a sneering voice, but impressed none the less, as I could easily see, by the precision of the other's manner.

'Is a rather curious person,' said Sherlock Holmes, turning upon his heel. 'I hope before very long to be able to introduce you to the pair of them. A word with you, Watson.'

He led me out to the head of the stair.

'This unexpected occurrence,' he said, 'has caused us rather to lose sight of the original purpose of our journey.'

'I have just been thinking so,' I answered; 'it is not right that Miss Morstan should remain in this stricken house.'

'No. You must escort her home. She lives with Mrs Cecil Forrester in Lower Camberwell, so it is not very far. I will wait for you here if you will drive out again. Or perhaps you are too tired?'

'By no means. I don't think I could rest until I know more of this fantastic business. I have seen something of the rough side of life, but I give you my word that this quick succession of strange surprises to-night has shaken my nerve completely. I should like, however, to see the matter through with you, now that I have got so far.'

'Your presence will be of great service to me,' he answered. 'We shall work the case out independently and leave this fellow Jones to exult over any mare's-nest* which he may choose to construct. When you have dropped Miss Morstan, I wish you to go on to No. 3, Pinchin Lane, down near the water's edge at Lambeth. The third house on the right-hand side is a bird-stuffer's; Sherman is the name. You will see a weasel holding a young rabbit in the window. Knock old Sherman up, and tell him, with my compliments, that I want Toby* at once. You will bring Toby back in the cab with you.'

'A dog, I suppose.'

'Yes, a queer mongrel, with a most amazing power of scent. I would rather have Toby's help than that of the whole detective force of London.'

'I shall bring him then,' said I. 'It is one now. I ought to be back before three, if I can get a fresh horse.'

'And I', said Holmes, 'shall see what I can learn from Mrs Bernstone and from the Indian servant, who, Mr Thaddeus tells me, sleeps in the next garret. Then I shall study the great Jones's methods and listen to his not too delicate sarcasms. "Wir sind gewohnt dass die Menschen verhöhnen was sie nicht verstehen."* Goethe is always pithy.'*

· CHAPTER 7 ·

The Episode of the Barrel

THE police had brought a cab with them, and in this I escorted Miss Morstan back to her home. After the angelic fashion of women, she had borne trouble with a calm face as long as there was someone weaker than herself to support, and I had found her bright and placid by the side of the frightened housekeeper. In the cab, however, she first turned faint, and then burst into a passion of weeping— so sorely had she been tried by the adventures of the night. She has told me since that she thought me cold and distant upon that journey. She little guessed the struggle within my breast, or the effort of self-restraint which held me back. My sympathies and my love went out to her, even as my hand had in the garden. I felt that years of the conventionalities of life could not teach me to know her sweet, brave nature

48

as had this one day of strange experiences. Yet there were two thoughts which sealed the words of affection upon my lips. She was weak and helpless, shaken in mind and nerve. It was to take her at a disadvantage to obtrude love upon her at such a time. Worse still, she was rich. If Holmes's researches were successful, she would be an heiress. Was it fair, was it honourable, that a half-pay surgeon* should take such advantage of an intimacy which chance had brought about? Might she not look upon me as a mere vulgar fortune-seeker? I could not bear to risk that such a thought should cross her mind. This Agra treasure intervened like an impassable barrier between us.

It was nearly two o'clock when we reached Mrs Cecil Forrester's. The servants had retired hours ago, but Mrs Forrester had been so interested by the strange message which Miss Morstan had received that she had sat up in the hope of her return. She opened the door herself, a middle-aged, graceful woman, and it gave me joy to see how tenderly her arm stole round the other's waist, and how motherly was the voice in which she greeted her. She was clearly no mere paid dependant, but an honoured friend. I was introduced, and Mrs Forrester earnestly begged me to step in and tell her our adventures. I explained, however, the importance of my errand, and promised faithfully to call and report any progress which we might make with the case. As we drove away I stole a glance back, and I still seem to see that little group on the step—the two graceful, clinging figures, the half-opened door, the hall-light shining through stained glass, the barometer, and the bright stair-rods. It was soothing to catch even that passing glimpse of a tranquil English home in the midst of the wild, dark business which had absorbed us.

And the more I thought of what had happened, the wilder and darker it grew. I reviewed the whole extraordinary sequence of events as I rattled on through the silent, gas-lit streets. There was the original problem: that at least was pretty clear now. The death of Captain Morstan, the sending of the pearls, the advertisement, the letter—we had

had light upon all those events. They had only led us, however, to a deeper and far more tragic mystery. The Indian treasure, the curious plan found among Morstan's baggage, the strange scene at Major Sholto's death, the rediscovery of the treasure immediately followed by the murder of the discoverer, the very singular accompaniments to the crime, the footsteps, the remarkable weapons, the words upon the card, corresponding with those upon Captain Morstan's chart—here was indeed a labyrinth in which a man less singularly endowed than my fellow-lodger might well despair of ever finding the clue.

Pinchin Lane was a row of shabby, two-storied brick houses in the lower quarter of Lambeth. I had to knock for some time at No. 3 before I could make any impression. At last, however, there was the glint of a candle behind the blind, and a face looked out at the upper window.

'Go on, you drunken vagabone,' said the face. 'If you kick up any more row, I'll open the kennels and let out forty-three dogs upon you.'

'If you'll let one out, it's just what I have come for,' said I.

'Go on!' yelled the voice. 'So help me gracious,* I have a wiper* in this bag, an' I'll drop it on your 'ead if you don't hook it!'

'But I want a dog,' I cried.

'I won't be argued with!' shouted Mr Sherman. 'Now stand clear; for when I say "three", down goes the wiper.'

'Mr Sherlock Holmes—' I began; but the words had a most magical effect, for the window instantly slammed down, and within a minute the door was unbarred and open. Mr Sherman was a lanky, lean old man, with stooping shoulders, a stringy neck, and blue-tinted glasses.

'A friend of Mr Sherlock is always welcome,' said he. 'Step in, sir. Keep clear of the badger, for he bites. Ah, naughty, naughty! would you take a nip at the gentleman?' This to a stoat which thrust its wicked head and red eyes between the bars of its cage. 'Don't mind that, sir; it's only a slowworm.* It hain't got no fangs, so I gives it the run o'

the room, for it keeps the beetles down. You must not mind
my bein' just a little short wi' you at first, for I'm guyed at*
by the children, and there's many a one just comes down
this lane to knock me up. What was it that Mr Sherlock
Holmes wanted, sir?'

'He wanted a dog of yours.'

'Ah! that would be Toby.'

'Yes, Toby was the name.'

'Toby lives at No. 7 on the left here.'

He moved slowly forward with his candle among the
queer animal family which he had gathered round him. In
the uncertain, shadowy light I could see dimly that there
were glancing, glimmering eyes peeping down at us from
every cranny and corner. Even the rafters above our heads
were lined by solemn fowls, who lazily shifted their weight
from one leg to the other as our voices disturbed their
slumbers.

Toby proved to be an ugly, long-haired, lop-eared creature, half spaniel and half lurcher,* brown and white in
colour, with a very clumsy, waddling gait. It accepted, after
some hesitation, a lump of sugar which the old naturalist
handed to me, and, having thus sealed an alliance, it
followed me to the cab, and made no difficulties about
accompanying me. It had just struck three on the Palace
clock when I found myself back once more at Pondicherry
Lodge. The ex-prize-fighter McMurdo had, I found, been
arrested as an accessory, and both he and Mr Sholto had
been marched off to the station. Two constables guarded the
narrow gate, but they allowed me to pass with the dog on
my mentioning the detective's name.

Holmes was standing on the doorstep with his hands in
his pockets, smoking his pipe.

'Ah, you have him there!' said he. 'Good dog, then!
Athelney Jones has gone. We have had an immense display
of energy since you left. He has arrested not only friend
Thaddeus but the gatekeeper, the housekeeper, and the
Indian servant. We have the place to ourselves but for a
sergeant upstairs. Leave the dog here, and come up.'

We tied Toby to the hall table, and reascended the stairs. The room was as we had left it, save that a sheet had been draped over the central figure. A weary-looking police-sergeant reclined in the corner.

'Lend me your bull's-eye,* sergeant,' said my companion. 'Now tie this bit of cord* round my neck, so as to hang it in front of me. Thank you. Now I must kick off my boots and stockings. Just you carry them down with you, Watson. I am going to do a little climbing. And dip my handkerchief into the creosote. That will do. Now come up into the garret with me for a moment.'

We clambered up through the hole. Holmes turned his light once more upon the footsteps in the dust.

'I wish you particularly to notice these foot-marks,' he said. 'Do you observe anything noteworthy about them?'

'They belong,' I said, 'to a child or a small woman.'

'Apart from their size, though. Is there nothing else?'

'They appear to be much as other foot-marks.'

'Not at all. Look here! This is the print of a right foot in the dust. Now I make one with my naked foot beside it. What is the chief difference?'

'Your toes are all cramped together. The other print has each toe distinctly divided.'

'Quite so. That is the point. Bear that in mind. Now, would you kindly step over to that flap-window and smell the edge of the wood-work? I shall stay over here, as I have this handkerchief in my hand.'

I did as he directed, and was instantly conscious of a strong tarry smell.

'That is where he put his foot in getting out. If *you* can trace him, I should think that Toby will have no difficulty. Now run downstairs, loose the dog, and look out for Blondin.'*

By the time that I got out into the grounds Sherlock Holmes was on the roof, and I could see him like an enormous glow-worm crawling very slowly along the ridge. I lost sight of him behind a stack of chimneys, but he presently reappeared and then vanished once more upon the opposite side. When I made my way round there I found him seated at one of the corner eaves.

'That you, Watson?' he cried.

'Yes.'

'This is the place. What is that black thing down there?'

'A water-barrel.'

'Top on it?'

'Yes.'

'No sign of a ladder?'

'No.'

'Confound the fellow! It's a most breakneck place. I ought to be able to come down where he could climb up. The water-pipe feels pretty firm. Here goes, anyhow.'

There was a scuffling of feet, and the lantern began to come steadily down the side of the wall. Then with a light spring he came on to the barrel, and from there to the earth.

'It was easy to follow him,' he said, drawing on his stockings and boots. 'Tiles were loosened the whole way along, and in his hurry he had dropped this. It confirms my diagnosis, as you doctors express it.'

The object which he held up to me was a small pocket* or pouch woven out of coloured grasses and with a few tawdry beads strung round it. In shape and size it was not unlike a cigarette-case. Inside were half a dozen spines of dark wood, sharp at one end and rounded at the other, like that which had struck Bartholomew Sholto.

'They are hellish things,' said he. 'Look out that you don't prick yourself. I'm delighted to have them, for the chances are that they are all he has. There is the less fear of you or me finding one in our skin before long. I would sooner face a Martini bullet,* myself. Are you game for a six-mile trudge, Watson?'

'Certainly,' I answered.

'Your leg will stand it?'

'Oh, yes.'

'Here you are, doggy! Good old Toby! Smell it, Toby, smell it!' He pushed the creosote handkerchief under the dog's nose, while the creature stood with its fluffy legs

53

separated, and with a most comical cock to its head, like a connoisseur sniffing the *bouquet* of a famous vintage. Holmes then threw the handkerchief to a distance, fastened a stout cord to the mongrel's collar, and led him to the foot of the water-barrel. The creature instantly broke into a succession of high, tremulous yelps, and, with his nose on the ground, and his tail in the air, pattered off upon the trail at a pace which strained his leash and kept us at the top of our speed.

The east had been gradually whitening, and we could now see some distance in the cold gray light. The square, massive house, with its black, empty windows and high, bare walls, towered up, sad and forlorn, behind us. Our course led right across the grounds, in and out among the trenches and pits with which they were scarred and intersected. The whole place, with its scattered dirt-heaps and ill-grown shrubs, had a blighted, ill-omened look which harmonized with the black tragedy which hung over it.

On reaching the boundary wall Toby ran along, whining eagerly, underneath its shadow, and stopped finally in a corner screened by a young beech. Where the two walls joined, several bricks had been loosened, and the crevices left were worn down and rounded upon the lower side, as though they had frequently been used as a ladder. Holmes clambered up, and, taking the dog from me, he dropped it over upon the other side.

'There's the print of wooden-leg's hand,' he remarked, as I mounted up beside him. 'You see the slight smudge of blood upon the white plaster. What a lucky thing it is that we have had no very heavy rain since yesterday! The scent will lie upon the road in spite of their eight-and-twenty hours' start.'

I confess that I had my doubts myself when I reflected upon the great traffic which had passed along the London road in the interval. My fears were soon appeased, however. Toby never hesitated or swerved, but waddled on in his peculiar rolling fashion. Clearly, the pungent smell of the creosote rose high above all other contending scents.

'Do not imagine,' said Holmes, 'that I depend for my success in this case upon the mere chance of one of these fellows having put his foot in the chemical. I have knowledge now which would enable me to trace them in many different ways. This, however, is the readiest, and, since fortune has put it into our hands, I should be culpable if I neglected it. It has, however, prevented the case from becoming the pretty little intellectual problem which it at one time promised to be. There might have been some credit to be gained out of it, but for this too palpable clue.'

'There is credit, and to spare,' said I. 'I assure you, Holmes, that I marvel at the means by which you obtain your results in this case even more than I did in the Jefferson Hope murder. The thing seems to me to be deeper and more inexplicable. How, for example, could you describe with such confidence the wooden-legged man?'

'Pshaw, my dear boy! it was simplicity itself. I don't wish to be theatrical. It is all patent and above-board. Two officers who are in command of a convict-guard learn an important secret as to buried treasure. A map is drawn for them by an Englishman named Jonathan Small. You remember that we saw the name upon the chart in Captain Morstan's possession. He had signed it in behalf of himself and his associates—the sign of the four, as he somewhat dramatically called it. Aided by this chart, the officers—or one of them—gets the treasure and brings it to England, leaving, we will suppose, some condition under which he received it unfulfilled. Now, then, why did not Jonathan Small get the treasure himself? The answer is obvious. The chart is dated at a time when Morstan was brought into close association with convicts. Jonathan Small did not get the treasure because he and his associates were themselves convicts and could not get away.'

'But this is mere speculation,' said I.

'It is more than that. It is the only hypothesis which covers the facts. Let us see how it fits in with the sequel. Major Sholto remains at peace for some years, happy in the possession of his treasure. Then he receives a letter from India which gives him a great fright. What was that?'

'A letter to say that the men whom he had wronged had been set free.'

'Or had escaped. That is much more likely, for he would have known what their term of imprisonment was. It would not have been a surprise to him. What does he do then? He guards himself against a wooden-legged man—a white man, mark you, for he mistakes a white tradesman for him, and actually fires a pistol at him. Now, only one white man's name is on the chart. The others are Hindoos or Mohammedans.* There is no other white man. Therefore we may say with confidence that the wooden-legged man is identical with Jonathan Small. Does the reasoning strike you as being faulty?'

'No: it is clear and concise.'

'Well, now, let us put ourselves in the place of Jonathan Small. Let us look at it from his point of view. He comes to England with the double idea of regaining what he would consider to be his rights and of having his revenge upon the man who had wronged him. He found out where Sholto lived, and very possibly he established communications with someone inside the house. There is this butler, Lal Rao, whom we have not seen. Mrs Bernstone gives him far from a good character. Small could not find out, however, where the treasure was hid, for no one ever knew, save the major and one faithful servant who had died. Suddenly Small learns that the major is on his death-bed. In a frenzy lest the secret of the treasure die with him, he runs the gauntlet of the guards, makes his way to the dying man's window, and is only deterred from entering by the presence of his two sons. Mad with hate, however, against the dead man, he enters the room that night, searches his private papers in the hope of discovering some memorandum relating to the treasure, and finally leaves a memento of his visit in the short inscription upon the card. He had doubtless planned beforehand that, should he slay the major, he would leave some such record upon the body as a sign that it was not a common murder, but, from the point of view of the four associates, something in the nature of an act of justice. Whimsical and bizarre conceits of this kind are common enough in the annals of crime, and usually afford valuable indications as to the criminal. Do you follow all this?'

'Very clearly.'

'Now, what could Jonathan Small do? He could only continue to keep a secret watch upon the efforts made to find the treasure. Possibly he leaves England and only comes back at intervals. Then comes the discovery of the garret, and he is instantly informed of it. We again trace the presence of some confederate in the household. Jonathan, with his wooden leg, is utterly unable to reach the lofty room of Bartholomew Sholto. He takes with him, however, a rather curious associate, who gets over this difficulty, but dips his naked foot into creosote, whence come Toby, and a six-mile limp for a half-pay officer with a damaged tendo Achillis.'*

'But it was the associate, and not Jonathan, who committed the crime.'

'Quite so. And rather to Jonathan's disgust, to judge by the way he stamped about when he got into the room. He bore no grudge against Bartholomew Sholto, and would have preferred if he could have been simply bound and gagged. He did not wish to put his head in a halter. There was no help for it, however: the savage instincts of his companion had broken out, and the poison had done its work: so Jonathan Small left his record, lowered the treasure-box to the ground, and followed it himself. That was the train of events as far as I can decipher them. Of course, as to his personal appearance, he must be middle-aged and must be sunburned after serving his time in such an oven as the Andamans. His height is readily calculated from the length of his stride, and we know that he was bearded. His hairiness was the one point which impressed itself upon Thaddeus Sholto when he saw him at the window. I don't know that there is anything else.'

'The associate?'

'Ah, well, there is no great mystery in that. But you will know all about it soon enough. How sweet the morning air is! See how that one little cloud floats like a pink feather from some gigantic flamingo. Now the red rim of the sun pushes itself over the London cloud-bank. It shines on a good many folk, but on none, I dare bet, who are on a stranger errand than you and I. How small we feel with our petty ambitions and strivings in the presence of the great elemental forces of Nature! Are you well up in your Jean Paul?'*

'Fairly so. I worked back to him through Carlyle.'*

'That was like following the brook to the parent lake. He makes one curious but profound remark. It is that the chief proof of man's real greatness lies in his perception of his own smallness. It argues, you see, a power of comparison and of appreciation which is in itself a proof of nobility. There is much food for thought in Richter. You have not a pistol, have you?'

'I have my stick.'

'It is just possible that we may need something of the sort if we get to their lair. Jonathan I shall leave to you, but if the other turns nasty I shall shoot him dead.'

He took out his revolver as he spoke, and, having loaded two of the chambers, he put it back into the right-hand pocket of his jacket.

We had during this time been following the guidance of Toby down the half-rural villa-lined roads which lead to the metropolis. Now, however, we were beginning to come among continuous streets, where labourers and dockmen were already astir, and slatternly women were taking down shutters and brushing door-steps. At the square-topped corner public-houses business was just beginning, and rough-looking men were emerging, rubbing their sleeves across their beards after their morning wet. Strange dogs sauntered up and stared wonderingly at us as we passed, but our inimitable Toby looked neither to the right nor to the left, but trotted onward with his nose to the ground and an occasional eager whine which spoke of a hot scent.

We had traversed Streatham, Brixton, Camberwell, and now found ourselves in Kennington Lane, having borne away through the side streets to the east of the Oval. The men whom we pursued seemed to have taken a curiously zigzag road, with the idea probably of escaping observation. They had never kept to the main road if a parallel side-street would serve their turn. At the foot of Kennington Lane they had edged away to the left through Bond Street and Miles Street. Where the latter street turns into Knight's Place, Toby ceased to advance but began to run backward and forward with one ear cocked and the other drooping, the very picture of canine indecision. Then he waddled round in circles, looking up to us from time to time, as if to ask for sympathy in his embarrassment.

'What the deuce is the matter with the dog?' growled Holmes. 'They surely would not take a cab, or go off in a balloon.'

'Perhaps they stood here for some time,' I suggested.

'Ah! it's all right. He's off again,' said my companion, in a tone of relief.

He was indeed off, for after sniffing round again he suddenly made up his mind, and darted away with an energy and determination such as he had not yet shown. The scent appeared to be much hotter than before, for he had not even to put his nose on the ground, but tugged at his leash and tried to break into a run. I could see by the gleam in Holmes's eyes that he thought we were nearing the end of our journey.

Our course now ran down Nine Elms until we came to Broderick and Nelson's large timber-yard, just past the White Eagle tavern. Here the dog, frantic with excitement, turned down through the side gate into the enclosure, where the sawyers were already at work. On the dog raced through sawdust and shavings, down an alley, round a passage, between two wood-piles, and finally, with a triumphant yelp, sprang upon a large barrel which still stood upon the hand-trolley on which it had been brought. With lolling tongue and blinking eyes, Toby stood upon the cask, looking from one to the other of us for some sign of appreciation. The staves of the barrel and the wheels of the trolley were smeared with a dark liquid, and the whole air was heavy with the smell of creosote.

Sherlock Holmes and I looked blankly at each other, and then burst simultaneously into an uncontrollable fit of laughter.

· CHAPTER 8 ·

The Baker Street Irregulars

'WHAT now?' I asked. 'Toby has lost his character for infallibility.'

'He acted according to his lights,' said Holmes, lifting him down from the barrel and walking him out of the timber-yard. 'If you consider how much creosote is carted about London in one day, it is no great wonder that our trail should have been crossed. It is much used now, especially for the seasoning of wood. Poor Toby is not to blame.'

'We must get on the main scent again, I suppose.'

'Yes. And, fortunately, we have no distance to go. Evidently what puzzled the dog at the corner of Knight's Place was that there were two different trails running in opposite directions. We took the wrong one. It only remains to follow the other.'

There was no difficulty about this. On leading Toby to the place where he had committed his fault, he cast about in a wide circle and finally dashed off in a fresh direction.

'We must take care that he does not now bring us to the place where the creosote-barrel came from,' I observed.

'I had thought of that. But you notice that he keeps on the pavement, whereas the barrel passed down the roadway. No, we are on the true scent now.'

It tended down towards the river-side, running through Belmont Place and Prince's Street. At the end of Broad Street it ran right down to the water's edge, where there was a small wooden wharf. Toby led us to the very edge of this, and there stood whining, looking out on the dark current beyond.

'We are out of luck,' said Holmes. 'They have taken to a boat here.'

Several small punts and skiffs were lying about in the water and on the edge of the wharf. We took Toby round to each in turn, but, though he sniffed earnestly, he made no sign.

Close to the rude landing-stage was a small brick house, with a wooden placard slung out through the second window. 'Mordecai Smith' was printed across it in large letters, and, underneath, 'Boats to hire by the hour or day'. A second inscription above the door informed us that a steam launch was kept—a statement which was confirmed by a great pile of coke upon the jetty. Sherlock Holmes looked slowly round, and his face assumed an ominous expression.

'This looks bad,' said he. 'These fellows are sharper than I expected. They seem to have covered their tracks. There has, I fear, been preconcerted management here.'

He was approaching the door of the house, when it opened, and a little curly-headed lad of six came running out, followed by a stoutish, red-faced woman with a large sponge in her hand.

'You come back and be washed, Jack,' she shouted. 'Come back, you young imp; for if your father comes home and finds you like that, he'll let us hear of it.'

'Dear little chap!' said Holmes strategically. 'What a rosy-cheeked young rascal! Now, Jack, is there anything you would like?'

The youth pondered for a moment.

'I'd like a shillin','* said he.

'Nothing you would like better?'

'I'd like two shillin' better,' the prodigy answered, after some thought.

'Here you are, then! Catch!—A fine child, Mrs Smith!'

'Lor' bless you, sir, he is that, and forward. He gets a'most too much for me to manage, 'specially when my man is away days at a time.'

'Away, is he?' said Holmes, in a disappointed voice. 'I am sorry for that, for I wanted to speak to Mr Smith.'

'He's been away since yesterday mornin', sir, and, truth to tell, I am beginnin' to feel frightened about him. But if it was about a boat, sir, maybe I could serve as well.'

'I wanted to hire his steam launch.'

'Why, bless you, sir, it is in the steam launch that he has gone. That's what puzzles me; for I know there ain't more coals in her than would take her to about Woolwich and back. If he'd been away in the barge I'd ha' thought nothin'; for many a time a job has taken him as far as Gravesend, and then if there was much doin' there he might ha' stayed over. But what good is a steam launch without coals?'

'He might have bought some at a wharf down the river.'

'He might, sir, but it weren't his way. Many a time I've heard him call out at the prices they charge for a few odd bags. Besides, I don't like that wooden-legged man, wi' his ugly face and outlandish talk. What did he want always knockin' about here for?'

'A wooden-legged man?' said Holmes with bland surprise.

'Yes, sir, a brown, monkey-faced chap that's called more'n once for my old man. It was him that roused him up yesternight, and, what's more, my man knew he was comin', for he had steam up in the launch. I tell you straight, sir, I don't feel easy in my mind about it.'

'But, my dear Mrs Smith,' said Holmes, shrugging his shoulders, 'you are frightening yourself about nothing. How could you possibly tell that it was the wooden-legged man who came in the night? I don't quite understand how you can be so sure.'

'His voice, sir. I knew his voice, which is kind o' thick and foggy. He tapped at the winder—about three it would be. "Show a leg, matey," says he: "time to turn out guard." My old man woke up Jim—that's my eldest—and away they went without so much as a word to me. I could hear the wooden leg clackin' on the stones.'

'And was this wooden-legged man alone?'

'Couldn't say, I am sure, sir. I didn't hear no one else.'

'I am sorry, Mrs Smith, for I wanted a steam launch, and I have heard good reports of the—Let me see, what is her name?'

'The *Aurora*, sir.'

'Ah! She's not that old green launch with a yellow line, very broad in the beam?'

'No, indeed. She's as trim a little thing as any on the river. She's been fresh painted, black with two red streaks.'

'Thanks. I hope that you will hear soon from Mr Smith. I am going down the river, and if I should see anything of the *Aurora* I shall let him know that you are uneasy. A black funnel, you say?'

'No, sir. Black with a white band.'

'Ah, of course. It was the sides which were black. Good morning, Mrs Smith. There is a boatman here with a wherry,* Watson. We shall take it and cross the river.'

'The main thing with people of that sort', said Holmes, as we sat in the sheets of the wherry,* 'is never to let them think that their information can be of the slightest importance to you. If you do, they will instantly shut up like an oyster. If you listen to them under protest, as it were, you are very likely to get what you want.'

'Our course now seems pretty clear,' said I.

'What would you do, then?'

'I would engage a launch and go down the river on the track of the *Aurora*.'

'My dear fellow, it would be a colossal task. She may have touched at any wharf on either side of the stream between here and Greenwich. Below the bridge there is a perfect labyrinth of landing-places for miles. It would take you days and days to exhaust them, if you set about it alone.'

'Employ the police, then.'

'No. I shall probably call Athelney Jones in at the last moment. He is not a bad fellow, and I should not like to do anything which would injure him professionally. But I have a fancy for working it out myself, now that we have gone so far.'

'Could we advertise, then, asking for information from wharfingers?'*

'Worse and worse! Our men would know that the chase was hot at their heels, and they would be off out of the country. As it is, they are likely enough to leave, but as long as they think they are perfectly safe they will be in no hurry. Jones's energy will be of use to us there, for his view of the case is sure to push itself into the daily press, and the runaways will think that everyone is off on the wrong scent.'

'What are we to do, then?' I asked as we landed near Millbank Penitentiary.*

'Take this hansom, drive home, have some breakfast, and get an hour's sleep. It is quite on the cards that we may be afoot to-night again. Stop at a telegraph office, cabby! We will keep Toby, for he may be of use to us yet.'

We pulled up at the Great Peter Street post-office, and Holmes despatched his wire.

'Whom do you think that is to?' he asked as we resumed our journey.

'I am sure I don't know.'

'You remember the Baker Street division of the detective police force whom I employed in the Jefferson Hope case?'

'Well,' said I, laughing.

'This is just the case where they might be invaluable. If they fail I have other resources, but I shall try them first. That wire was to my dirty little lieutenant, Wiggins, and I expect that he and his gang will be with us before we have finished our breakfast.'

It was between eight and nine o'clock now, and I was conscious of a strong reaction after the successive excitements of the night. I was limp and weary, befogged in mind and fatigued in body. I had not the professional enthusiasm which carried my companion on, nor could I look at the matter as a mere abstract intellectual problem. As far as the death of Bartholomew Sholto went, I had heard little good of him, and could feel no intense antipathy to his murderers. The treasure, however, was a different matter. That, or part of it, belonged rightfully to Miss Morstan. While there was a chance of recovering it, I was ready to devote my life to the one object. True, if I found it, it would probably put her forever beyond my reach. Yet it would be a petty and selfish love which would be influenced by such a thought as that. If Holmes could work to find the criminals, I had a tenfold stronger reason to urge me on to find the treasure.

A bath at Baker Street and a complete change freshened me up wonderfully. When I came down to our room I found the breakfast laid and Holmes pouring out the coffee.

'Here it is,' said he, laughing and pointing to an open newspaper. 'The energetic Jones and the ubiquitous reporter have fixed it up between them. But you have had enough of the case. Better have your ham and eggs first.'

I took the paper from him and read the short notice, which was headed 'Mysterious Business at Upper Norwood'.

About twelve o'clock last night [said the *Standard*] Mr Bartholomew Sholto, of Pondicherry Lodge, Upper Norwood, was found dead in his room under circumstances which point to foul play. As far as we can learn, no actual traces of violence were found upon Mr Sholto's person, but a valuable collection of Indian gems which the deceased gentleman had inherited from his father has been carried off. The discovery was first made by Mr Sherlock Holmes and Dr Watson, who had called at the house with Mr Thaddeus Sholto, brother of the deceased. By a singular piece of good fortune, Mr Athelney Jones, the well-known member of the detective police force, happened to be at the Norwood Police Station, and was on the ground within half an hour of the first alarm. His trained and experienced faculties were at once directed towards the detection of the criminals, with the gratifying result that the brother, Thaddeus Sholto, has already been arrested, together with the housekeeper, Mrs Bernstone, an Indian butler named Lal Rao, and a porter, or gatekeeper, named McMurdo. It is quite certain that the thief or thieves were well acquainted with the house, for Mr Jones's well-known technical knowledge and his powers of minute observation have enabled him to prove conclusively that the miscreants could not have entered by the door or by the window, but must have made their way across the roof of the building, and so through a trap-door into a room which communicated with that in which the body was found. This fact, which has been very clearly made out, proves conclusively that it was no mere haphazard burglary. The prompt and energetic action of the officers of the law shows the great advantage of the presence on such occasions of a single vigorous and masterful mind. We cannot but think that it supplies an argument to those who would wish to see our detectives more decentralized, and so brought into closer and more effective touch with the cases which it is their duty to investigate.

'Isn't it gorgeous!' said Holmes, grinning over his coffee cup. 'What do you think of it?'

'I think that we have had a close shave ourselves of being arrested for the crime.'

'So do I. I wouldn't answer for our safety now, if he should happen to have another of his attacks of energy.'

At this moment there was a loud ring at the bell, and I could hear Mrs Hudson, our landlady, raising her voice in a wail of expostulation and dismay.

'By heavens, Holmes,' I said, half rising, 'I believe that they are really after us.'

'No, it's not quite so bad as that. It is the unofficial force—the Baker Street irregulars.'

As he spoke, there came a swift pattering of naked feet upon the stairs, a clatter of high voices, and in rushed a dozen dirty and ragged little street arabs. There was some show of discipline among them, despite their tumultuous entry, for they instantly drew up in line and stood facing us with expectant faces. One of their number, taller and older than the others, stood forward with an air of lounging superiority which was very funny in such a disreputable little scarecrow.

'Got your message, sir,' said he, 'and brought 'em on sharp. Three bob and a tanner* for tickets.'

'Here you are,' said Holmes, producing some silver. 'In future they can report to you, Wiggins, and you to me. I cannot have the house invaded in this way. However, it is just as well that you should all hear the instructions. I want to find the whereabouts of a steam launch called the *Aurora*, owner Mordecai Smith, black with two red streaks, funnel black with a white band. She is down the river somewhere. I want one boy to be at Mordecai Smith's landing-stage opposite Millbank to say if the boat comes back. You must divide it out among yourselves and do both banks thoroughly. Let me know the moment you have news. Is that all clear?'

'Yes, guv'nor,' said Wiggins.

'The old scale of pay, and a guinea to the boy who finds the boat. Here's a day in advance. Now off you go!'

He handed them a shilling each, and away they buzzed down the stairs, and I saw them a moment later streaming down the street.

'If the launch is above water they will find her,' said Holmes, as he rose from the table and lit his pipe. 'They can go everywhere, see everything, overhear everyone. I expect to hear before evening that they have spotted her. In the meanwhile, we can do nothing but await results. We cannot pick up the broken trail until we find either the *Aurora* or Mr Mordecai Smith.'

'Toby could eat these scraps, I dare say. Are you going to bed, Holmes?'

'No: I am not tired. I have a curious constitution. I never remember feeling tired by work, though idleness exhausts me completely. I am going to smoke and to think over this queer business to which my fair client has introduced us. If ever man had an easy task, this of ours ought to be. Wooden-legged men are not so common, but the other man must, I should think, be absolutely unique.'

'That other man again!'

'I have no wish to make a mystery of him to you, anyway. But you must have formed your own opinion. Now, do consider the data. Diminutive foot-marks, toes never fettered by boots, naked feet, stone-headed wooden mace, great agility, small poisoned darts. What do you make of all this?'

'A savage!' I exclaimed. 'Perhaps one of those Indians who were the associates of Jonathan Small.'

'Hardly that,' said he. 'When first I saw signs of strange weapons, I was inclined to think so; but the remarkable character of the foot-marks caused me to reconsider my views. Some of the inhabitants of the Indian Peninsula are small men, but none could have left such marks as that. The Hindoo proper has long and thin feet. The sandal-wearing Mohammedan has the great toe well separated from the others, because the thong is commonly passed between. These little darts, too, could only be shot in one way. They are from a blow-pipe. Now, then, where are we to find our savage?'

'South America,' I hazarded.

He stretched his hand up, and took down a bulky volume from the shelf.

'This is the first volume of a gazetteer which is now being published. It may be looked upon as the very latest authority. What have we here? "Andaman Islands, situated 340 miles to the north of Sumatra, in the Bay of Bengal." Hum! hum! What's all this? Moist climate, coral reefs, sharks, Port Blair,* convict barracks, Rutland Island, cotton-woods—Ah, here we are! "The aborigines of the Andaman Islands may perhaps claim the distinction of being the smallest race upon this earth, though some anthropologists prefer the Bushmen of Africa, the Digger Indians of America, and the Terra del Fuegians. The average height is rather below four feet, although many full-grown adults may be found who are very much smaller than this. They are a fierce, morose, and intractable people, though capable of forming most devoted friendships when their confidence has once been gained."* Mark that, Watson. Now, then listen to this. "They are naturally hideous, having large, misshapen heads, small fierce eyes, and distorted features. Their feet and hands, however, are remarkably small. So intractable and fierce are they, that all the efforts of the British officials have failed to win them over in any degree. They have always been a terror to shipwrecked crews, braining the survivors with their stone-headed clubs, or shooting them with their poisoned arrows. These massacres are invariably concluded by a cannibal feast." Nice, amiable people, Watson! If this fellow had been left to his own unaided devices, this affair might have taken an even more ghastly turn. I fancy that, even as it is, Jonathan Small would give a good deal not to have employed him.'

'But how came he to have so singular a companion?'

'Ah, that is more than I can tell. Since, however, we had already determined that Small had come from the Andamans, it is not so very wonderful that this islander should be with him. No doubt we shall know all about it in time. Look here, Watson; you look regularly done. Lie down there on the sofa, and see if I can put you to sleep.'

He took up his violin from the corner, and as I stretched myself out he began to play some low, dreamy, melodious air—his own, no doubt, for he had a remarkable gift for improvisation. I have a vague remembrance of his gaunt limbs, his earnest face, and the rise and fall of his bow. Then I seemed to be floated peacefully away upon a soft sea of sound, until I found myself in dreamland, with the sweet face of Mary Morstan looking down upon me.

· CHAPTER 9 ·

A Break in the Chain

IT was late in the afternoon before I woke, strengthened and refreshed. Sherlock Holmes still sat exactly as I had left him, save that he had laid aside his violin and was deep in a book. He looked across at me as I stirred, and I noticed that his face was dark and troubled.

'You have slept soundly,' he said. 'I feared that our talk would wake you.'

'I heard nothing,' I answered. 'Have you had fresh news, then?'

'Unfortunately, no. I confess that I am surprised and disappointed. I expected something definite by this time. Wiggins has just been up to report. He says that no trace can be found of the launch. It is a provoking check, for every hour is of importance.'

'Can I do anything? I am perfectly fresh now, and quite ready for another night's outing.'

'No; we can do nothing. We can only wait. If we go ourselves, the message might come in our absence and delay be caused. You can do what you will, but I must remain on guard.'

'Then I shall run over to Camberwell and call upon Mrs Cecil Forrester. She asked me to, yesterday.'

'On Mrs Cecil Forrester?' asked Holmes, with the twinkle of a smile in his eyes.

'Well, of course on Miss Morstan, too. They were anxious to hear what happened.'

'I would not tell them too much,' said Holmes. 'Women are never to be entirely trusted—not the best of them.'

I did not pause to argue over this atrocious sentiment.

'I shall be back in an hour or two,' I remarked.

'All right! Good luck! But, I say, if you are crossing the river you may as well return Toby, for I don't think it is at all likely that we shall have any use for him now.'

I took our mongrel accordingly and left him, together with a half-sovereign,* at the old naturalist's in Pinchin Lane. At Camberwell I found Miss Morstan a little weary after her night's adventures, but very eager to hear the news. Mrs Forrester, too, was full of curiosity. I told them all that we had done, suppressing, however, the more dreadful parts of the tragedy. Thus, although I spoke of Mr Sholto's death, I said nothing of the exact manner and method of it. With all my omissions, however, there was enough to startle and amaze them.

'It is a romance!' cried Mrs Forrester. 'An injured lady, half a million in treasure, a black cannibal, and a wooden-legged ruffian. They take the place of the conventional dragon or wicked earl.'

'And two knight-errants to the rescue,' added Miss Morstan, with a bright glance at me.

'Why, Mary, your fortune depends upon the issue of this search. I don't think that you are nearly excited enough. Just imagine what it must be to be so rich, and to have the world at your feet!'

It sent a little thrill of joy to my heart to notice that she showed no sign of elation at the prospect. On the contrary, she gave a toss of her proud head, as though the matter were one in which she took small interest.

'It is for Mr Thaddeus Sholto that I am anxious,' she said. 'Nothing else is of any consequence; but I think that he has behaved most kindly and honourably throughout. It is our duty to clear him of this dreadful and unfounded charge.'

It was evening before I left Camberwell, and quite dark by the time I reached home. My companion's book and pipe lay by his chair, but he had disappeared. I looked about in the hope of seeing a note, but there was none.

'I suppose that Mr Sherlock Holmes has gone out,' I said to Mrs Hudson as she came up to lower the blinds.

'No, sir. He has gone to his room, sir. Do you know, sir,' sinking her voice into an impressive whisper, 'I am afraid for his health?'

'Why so, Mrs Hudson?'

'Well, he's that strange, sir. After you was gone he walked and he walked, up and down, and up and down, until I was weary of the sound of his footstep. Then I heard him talking to himself and muttering, and every time the bell rang out he came on the stair-head, with "What is that, Mrs Hudson?" And now he has slammed off to his room, but I can hear him walking away the same as ever. I hope he's not going to be ill, sir. I ventured to say something to him about cooling medicine,* but he turned on me, sir, with such a look that I don't know how ever I got out of the room.'

'I don't think that you have any cause to be uneasy, Mrs Hudson,' I answered. 'I have seen him like this before. He has some small matter upon his mind which makes him restless.'

I tried to speak lightly to our worthy landlady, but I was myself somewhat uneasy when through the long night I still from time to time heard the dull sound of his tread, and knew how his keen spirit was chafing against this involuntary inaction.

At breakfast-time he looked worn and haggard, with a little fleck of feverish colour upon either cheek.

'You are knocking yourself up, old man,' I remarked. 'I heard you marching about in the night.'

'No, I could not sleep,' he answered. 'This infernal problem is consuming me. It is too much to be balked by so petty an obstacle, when all else had been overcome. I know the men, the launch, everything; and yet I can get no news. I have set other agencies at work, and used every means at my disposal. The whole river has been searched on either side, but there is no news, nor has Mrs Smith heard of her husband. I shall come to the conclusion soon that they have scuttled the craft. But there are objections to that.'

'Or that Mrs Smith has put us on a wrong scent.'

'No, I think that may be dismissed. I had inquiries made, and there is a launch of that description.'

'Could it have gone up the river?'

'I have considered that possibility, too, and there is a search-party who will work up as far as Richmond. If no news comes to-day I shall start off myself tomorrow and go for the men rather than the boat. But surely, surely, we shall hear something.'

We did not, however. Not a word came to us either from Wiggins or from the other agencies. There were articles in most of the papers upon the Norwood tragedy. They all appeared to be rather hostile to the unfortunate Thaddeus Sholto. No fresh details were to be found, however, in any of them, save that an inquest was to be held upon the following day. I walked over to Camberwell in the evening to report our ill-success to the ladies, and on my return I found Holmes dejected and somewhat morose. He would hardly reply to my questions, and busied himself all the evening in an abstruse chemical analysis which involved much heating of retorts and distilling of vapours, ending at last in a smell which fairly drove me out of the apartment. Up to the small hours of the morning I could hear the clinking of his test-tubes which told me that he was still engaged in his malodorous experiment.

In the early dawn I woke with a start and was surprised to find him standing by my bedside, clad in a rude sailor dress with a pea-jacket,* and a coarse red scarf round his neck.

'I am off down the river, Watson,' said he. 'I have been turning it over in my mind, and I can see only one way out of it. It is worth trying, at all events.'

'Surely I can come with you, then?' said I.

'No; you can be much more useful if you will remain here as my representative. I am loath to go, for it is quite on the cards that some message may come during the day, though Wiggins was despondent about it last night. I want you to open all notes and telegrams, and to act on your own judgment if any news should come. Can I rely upon you?'

'Most certainly.'

'I am afraid that you will not be able to wire me, for I can hardly tell yet where I may find myself. If I am in luck, however, I may not be gone so very long. I shall have news of some sort or other before I get back.'

I had heard nothing of him by breakfast-time. On opening the *Standard*, however, I found that there was a fresh allusion to the business.

With reference to the Upper Norwood tragedy [it remarked] we have reason to believe that the matter promises to be even more complex and mysterious than was originally supposed. Fresh evidence has shown that it is quite impossible that Mr Thaddeus Sholto could have been in any way concerned in the matter. He and the housekeeper, Mrs Bernstone, were both released yesterday evening. It is believed, however, that the police have a clue as to the real culprits, and that it is being prosecuted by Mr Athelney Jones, of Scotland Yard, with all his well-known energy and sagacity. Further arrests may be expected at any moment.

'That is satisfactory so far as it goes,' thought I. 'Friend Sholto is safe, at any rate. I wonder what the fresh clue may be, though it seems to be a stereotyped form whenever the police have made a blunder.'

I tossed the paper down upon the table, but at that moment my eye caught an advertisement in the agony column. It ran in this way:

LOST—Whereas Mordecai Smith, boatman, and his son Jim, left Smith's Wharf at or about three o'clock last Tuesday morning in the steam launch *Aurora*, black with two red stripes, funnel black with a white band, the sum of five pounds will be paid to anyone who can give information to Mrs Smith, at Smith's Wharf, or at 221B, Baker Street, as to the whereabouts of the said Mordecai Smith and the launch *Aurora*.

This was clearly Holmes's doing. The Baker Street address was enough to prove that. It struck me as rather ingenious, because it might be read by the fugitives without their seeing in it more than the natural anxiety of a wife for her missing husband.

It was a long day. Every time that a knock came to the door, or a sharp step passed in the street, I imagined that it was either Holmes returning or an answer to his advertisement. I tried to read, but my thoughts would wander off to our strange quest and to the ill-assorted and villainous pair whom we were pursuing. Could there be, I wondered, some radical flaw in my companion's reasoning? Might he not be suffering from some huge self-deception? Was it not possible that his nimble and speculative mind had built up this wild theory upon faulty premises? I had never known him to be wrong, and yet the keenest reasoner may occasionally be deceived. He was likely, I thought, to fall into error through the over-refinement of his logic—his preference for a subtle and bizarre explanation when a plainer and more commonplace one lay ready to his hand. Yet, on the other hand, I had myself seen the evidence, and I had heard the reasons for his deductions. When I looked back on the long chain of curious circumstances, many of them trivial in themselves, but all tending in the same direction, I could not disguise from myself that even if Holmes's explanation were incorrect the true theory must be equally *outré* * and startling.

At three o'clock on the afternoon there was a loud peal at the bell, an authoritative voice in the hall, and, to my surprise, no less a person than Mr Athelney Jones was shown up to me. Very different was he, however, from the brusque and masterful professor of common sense who had taken over the case so confidently at Upper Norwood. His expression was downcast, and his bearing meek and even apologetic.

'Good-day, sir; good-day,' said he. 'Mr Sherlock Holmes is out, I understand.'

'Yes, and I cannot be sure when he will be back. But perhaps you would care to wait. Take that chair and try one of these cigars.'

'Thank you; I don't mind if I do,' said he, mopping his face with a red bandanna* handkerchief.

'And a whisky and soda?'

'Well, half a glass. It is very hot for the time of year; and I have had a good deal to worry and try me. You know my theory about this Norwood case?'

'I remember that you expressed one.'

'Well, I have been obliged to reconsider it. I had my net drawn tightly round Mr Sholto, sir, when pop he went through a hole in the middle of it. He was able to prove an alibi which could not be shaken. From the time that he left his brother's room he was never out of sight of someone or other. So it could not be he who climbed over roofs and through trap-doors. It's a very dark case, and my professional credit is at stake. I should be very glad of a little assistance.'

'We all need help sometimes,' said I.

'Your friend, Mr Sherlock Holmes, is a wonderful man, sir,' said he, in a husky and confidential voice. 'He's a man who is not to be beat. I have known that young man go into a good many cases, but I never saw the case yet that he could not throw a light upon. He is irregular in his methods and a little quick perhaps in jumping at theories, but, on the whole, I think he would have made a most promising officer,* and I don't care who knows it. I have had a wire from him this morning, by which I understand that he has got some clue to this Sholto business. Here is his message.'

He took the telegram out of his pocket, and handed it to me. It was dated from Poplar at twelve o'clock.

Go to Baker Street at once [it said]. If I have not returned, wait for me. I am close on the track of the Sholto gang. You can come with us to-night if you want to be in at the finish.

'This sounds well. He has evidently picked up the scent again,' said I.

'Ah, then he has been at fault too,' exclaimed Jones with evident satisfaction. 'Even the best of us are thrown off sometimes. Of course this may prove to be a false alarm; but it is my duty as an officer of the law to allow no chance to slip. But there is someone at the door. Perhaps this is he.'

A heavy step was heard ascending the stair, with a great wheezing and rattling as from a man who was sorely put to it for breath. Once or twice he stopped, as though the climb were too much for him, but at last he made his way to our door and entered. His appearance corresponded to the sounds which we had heard. He was an aged man, clad in seafaring garb, with an old pea-jacket buttoned up to his throat. His back was bowed, his knees were shaky, and his breathing was painfully asthmatic. As he leaned upon a thick oaken cudgel his shoulders heaved in the effort to draw the air into his lungs. He had a coloured scarf round his chin, and I could see little of his face save a pair of keen dark eyes, overhung by bushy white brows, and long gray side-whiskers. Altogether he gave me the impression of a respectable master mariner who had fallen into years and poverty.

'What is it, my man?' I asked.

He looked about him in the slow methodical fashion of old age.

76

'Is Mr Sherlock Holmes here?' said he.

'No; but I am acting for him. You can tell me any message you have for him.'

'It was to him himself I was to tell it,' said he.

'But I tell you that I am acting for him. Was it about Mordecai Smith's boat?'

'Yes. I knows well where it is. An' I knows where the men he is after are. An' I knows where the treasure is. I knows all about it.'

'Then tell me, and I shall let him know.'

'It was to him I was to tell it,' he repeated with the petulant obstinacy of a very old man.

'Well, you must wait for him.'

'No, no; I ain't goin' to lose a whole day to please no one. If Mr Holmes ain't here, then Mr Holmes must find it all out for himself. I don't care about the look of either of you, and I won't tell a word.'

He shuffled towards the door, but Athelney Jones got in front of him.

'Wait a bit, my friend,' said he. 'You have important information, and you must not walk off. We shall keep you, whether you like it or not, until our friend returns.'

The old man made a little run towards the door, but, as Athelney Jones put his broad back up against it, he recognized the uselessness of resistance.

'Pretty sort o' treatment this!' he cried, stamping his stick. 'I come here to see a gentleman, and you two, who I never saw in my life, seize me and treat me in this fashion!'

'You will be none the worse,' I said. 'We shall recompense you for the loss of your time. Sit over here on the sofa, and you will not have long to wait.'

He came across sullenly enough and seated himself with his face resting on his hands. Jones and I resumed our cigars and our talk. Suddenly, however, Holmes's voice broke in upon us.

'I think that you might offer me a cigar too,' he said.

We both started in our chairs. There was Holmes sitting close to us with an air of quiet amusement.

77

'Holmes!' I exclaimed. 'You here! But where is the old man?'

'Here is the old man,' said he, holding out a heap of white hair. 'Here he is—wig, whiskers, eyebrows, and all. I thought my disguise was pretty good, but I hardly expected that it would stand that test.'*

'Ah, you rogue!' cried Jones, highly delighted. 'You would have made an actor and a rare one. You had the proper workhouse cough, and those weak legs of yours are worth ten pound a week. I thought I knew the glint of your eye, though. You didn't get away from us so easily, you see.'

'I have been working in that get-up all day,' said he, lighting his cigar. 'You see, a good many of the criminal classes begin to know me—especially since our friend here took to publishing some of my cases: so I can only go on the war-path under some simple disguise like this. You got my wire?'

'Yes; that was what brought me here.'

'How has your case prospered?'

'It has all come to nothing. I have had to release two of my prisoners, and there is no evidence against the other two.'

'Never mind. We shall give you two others in the place of them. But you must put yourself under my orders. You are welcome to all the official credit, but you must act on the lines that I point out. Is that agreed?'

'Entirely, if you will help me to the men.'

'Well, then, in the first place I shall want a fast police-boat—a steam launch—to be at the Westminster Stairs* at seven o'clock.'

'That is easily managed. There is always one about there; but I can step across the road and telephone to make sure.'

'Then I shall want two staunch men, in case of resistance.'

'There will be two or three in the boat. What else?'

'When we secure the men we shall get the treasure. I think that it would be a pleasure to my friend here to take the box round to the young lady to whom half of it rightfully belongs. Let her be the first to open it. Eh, Watson?'

'It would be a great pleasure to me.'

'Rather an irregular proceeding,' said Jones, shaking his head. 'However, the whole thing is irregular, and I suppose we must wink at it. The treasure must afterwards be handed over to the authorities until after the official investigation.'

'Certainly. That is easily managed. One other point. I should much like to have a few details about this matter from the lips of Jonathan Small himself. You know I like to work the details of my cases out. There is no objection to my having an unofficial interview with him, either here in my rooms or elsewhere, as long as he is efficiently guarded?'

'Well, you are master of the situation. I have had no proof yet of the existence of this Jonathan Small. However, if you can catch him, I don't see how I can refuse you an interview with him.'

'That is understood, then?'

'Perfectly. Is there anything else?'

'Only that I insist upon your dining with us. It will be ready in half an hour. I have oysters and a brace of grouse, with something a little choice in white wines.—Watson, you have never yet recognized my merits as a housekeeper.'

· CHAPTER 10 ·

The End of the Islander

OUR meal was a merry one. Holmes could talk exceedingly well when he chose, and that night he did choose. He appeared to be in a state of nervous exaltation. I have never known him so brilliant. He spoke on a quick succession of subjects—on miracle plays,* on medieval pottery, on Stradivarius violins, on the Buddhism of Ceylon, and on the warships of the future—handling each as though he had made a special study of it. His bright humour marked the reaction from his black depression of the preceding days. Athelney Jones proved to be a sociable soul in his hours of relaxation, and faced his dinner with the air of a *bon vivant.** For myself, I felt elated at the thought that we were nearing the end of our task, and I caught something of Holmes's gaiety. None of us alluded during dinner to the cause which had brought us together.

When the cloth was cleared Holmes glanced at his watch, and filled up three glasses with port.

'One bumper,'* said he, 'to the success of our little expedition. And now it is high time we were off. Have you a pistol, Watson?'

'I have my old service-revolver in my desk.'

'You had best take it, then. It is well to be prepared. I see that the cab is at the door. I ordered it for half-past six.'

It was a little past seven before we reached the Westminster wharf and found our launch awaiting us. Holmes eyed it critically.

'Is there anything to mark it as a police-boat?'

'Yes, that green lamp at the side.'

'Then take it off.'

The small change was made, we stepped on board, and the ropes were cast off. Jones, Holmes, and I sat in the stern. There was one man at the rudder, one to tend the engines, and two burly police-inspectors forward.

'Where to?' asked Jones.

'To the Tower.* Tell them to stop opposite to Jacobson's Yard.'

Our craft was evidently a very fast one. We shot past the long lines of loaded barges as though they were stationary. Holmes smiled with satisfaction as we overhauled a river steamer and left her behind us.

'We ought to be able to catch anything on the river,' he said.

'Well, hardly that. But there are not many launches to beat us.'

'We shall have to catch the *Aurora*, and she has a name for being a clipper. I will tell you how the land lies, Watson. You recollect how annoyed I was at being baulked by so small a thing?'

'Yes.'

'Well, I gave my mind a thorough rest by plunging into a chemical analysis. One of our greatest statesmen has said that a change of work is the best rest.* So it is. When I had succeeded in dissolving the hydrocarbon which I was at work at, I came back to our problem of the Sholtos, and thought the whole matter out again. My boys had been up the river and down the river without result. The launch was not at any landing-stage or wharf, nor had it returned. Yet it could hardly have been scuttled to hide their traces, though that always remained as a possible hypothesis if all else failed. I knew that this man Small had a certain degree of low cunning, but I did not think him capable of anything in the nature of delicate finesse. That is usually a product of higher education. I then reflected that since he had certainly been in London some time—as we had evidence that he maintained a continual watch over Pondicherry Lodge—he could hardly leave at a moment's notice, but would need some little time, if it were only a day, to arrange his affairs. That was the balance of probability, at any rate.'

'It seems to me to be a little weak,' said I; 'it is more probable that he had arranged his affairs before ever he set out upon his expedition.'

'No, I hardly think so. This lair of his would be too valuable a retreat in case of need for him to give it up until he was sure that he could do without it. But a second consideration struck me. Jonathan Small must have felt that the peculiar appearance of his companion, however much he may have top-coated him, would give rise to gossip, and possibly be associated with this Norwood tragedy. He was quite sharp enough to see that. They had started from their headquarters under cover of darkness, and he would wish to get back before it was broad light. Now, it was past three o'clock, according to Mrs Smith, when they got the boat. It would be quite bright, and people would be about in an hour or so. Therefore, I argued, they did not go very far.

They paid Smith well to hold his tongue, reserved his launch for the final escape, and hurried to their lodgings with the treasure-box. In a couple of nights, when they had time to see what view the papers took, and whether there was any suspicion, they would make their way under cover of darkness to some ship at Gravesend or in the Downs,* where no doubt they had already arranged for passages to America or the Colonies.'

'But the launch? They could not have taken that to their lodgings.'

'Quite so. I argued that the launch must be no great way off, in spite of its invisibility. I then put myself in the place of Small and looked at it as a man of his capacity would. He would probably consider that to send back the launch or to keep it at a wharf would make pursuit easy if the police did happen to get on his track. How, then, could he conceal the launch and yet have her at hand when wanted? I wondered what I should do myself if I were in his shoes. I could only think of one way of doing it. I might hand the launch over to some boat-builder or repairer, with directions to make a trifling change in her. She would then be removed to his shed or yard, and so be effectually concealed, while at the same time I could have her at a few hours' notice.'

'That seems simple enough.'

'It is just these very simple things which are extremely liable to be overlooked. However, I determined to act on the idea. I started at once in this harmless seaman's rig and enquired at all the yards down the river. I drew blank at fifteen, but at the sixteenth—Jacobson's—I learned that the *Aurora* had been handed over to them two days ago by a wooden-legged man, with some trivial directions as to her rudder. "There ain't naught amiss with her rudder," said the foreman. "There she lies, with the red streaks." At that moment who should come down but Mordecai Smith, the missing owner. He was rather the worse for liquor. I should not, of course, have known him, but he bellowed out his name and the name of his launch. "I want her tonight at eight o'clock," said he—"eight o'clock sharp, mind, for I have two gentlemen who won't be kept waiting." They had evidently paid him well, for he was very flush of money, chucking shillings about to the men. I followed him some distance, but he subsided into an alehouse; so I went back to the yard, and, happening to pick up one of my boys on the way, I stationed him as a sentry over the launch. He is to stand at the water's edge and wave his handkerchief to us when they start. We shall be lying off in the stream, and it will be a strange thing if we do not take men, treasure, and all.'

'You have planned it all very neatly, whether they are the right men or not,' said Jones; 'but if the affair were in my hands I should have had a body of police in Jacobson's Yard, and arrested them when they came down.'

'Which would have been never. This man Small is a pretty shrewd fellow. He would send a scout on ahead, and if anything made him suspicious he would lie snug for another week.'

'But you might have stuck to Mordecai Smith, and so been led to their hiding-place,' said I.

'In that case I should have wasted my day. I think that it is a hundred to one against Smith knowing where they live. As long as he has liquor and good pay, why should he ask questions? They send him messages what to do. No, I thought over every possible course, and this is the best.'

While this conversation had been proceeding, we had been shooting the long series of bridges which span the Thames. As we passed the City the last rays of the sun were gilding the cross upon the summit of St Paul's. It was twilight before we reached the Tower.

'That is Jacobson's Yard,' said Holmes, pointing to a bristle of masts and rigging on the Surrey side. 'Cruise gently up and down here under cover of this string of lighters.' He took a pair of night-glasses from his pocket and gazed some time at the shore. 'I see my sentry at his post,' he remarked, 'but no sign of a handkerchief.'

'Suppose we go downstream a short way and lie in wait for them,' said Jones eagerly.

We were all eager by this time, even the policemen and stokers, who had a very vague idea of what was going forward.

'We have no right to take anything for granted,' Holmes answered. 'It is certainly ten to one that they go downstream, but we cannot be certain. From this point we can see the entrance of the yard, and they can hardly see us. It will be a clear night and plenty of light. We must stay where we are. See how the folk swarm over yonder in the gaslight.'

'They are coming from work in the yard.'

'Dirty-looking rascals, but I suppose every one has some little immortal spark concealed about him. You would not think it, to look at them. There is no *a priori** probability about it. A strange enigma is man!'

'Someone calls him a soul concealed in an animal,' I suggested.

'Winwood Reade is good upon the subject,' said Holmes. 'He remarks that, while the individual man is an insoluble puzzle, in the aggregate he becomes a mathematical certainty. You can, for example, never foretell what any one man will do, but you can say with precision what an average number will be up to. Individuals vary, but percentages remain constant. So says the statistician. But do I see a handkerchief? Surely there is a white flutter over yonder.'

'Yes, it is your boy,' I cried. 'I can see him plainly.'

'And there is the *Aurora*,' exclaimed Holmes, 'and going like the devil! Full speed ahead, engineer. Make after that launch with the yellow light. By heaven, I shall never forgive myself if she proves to have the heels of us!'

She had slipped unseen through the yard-entrance, and passed between two or three small craft, so that she had fairly got her speed up before we saw her. Now she was flying down the stream, near in to the shore, going at a tremendous rate. Jones looked gravely at her and shook his head.

'She is very fast,' he said. 'I doubt if we shall catch her.'

'We *must* catch her!' cried Holmes, between his teeth. 'Heap it on, stokers! Make her do all she can! If we burn the boat we must have them!'

84

We were fairly after her now. The furnaces roared, and the powerful engines whizzed and clanked, like a great metallic heart. Her sharp, steep prow cut through the still river-water and sent two rolling waves to right and to left of us. With every throb of the engines we sprang and quivered like a living thing. One great yellow lantern in our bows threw a long, flickering funnel of light in front of us. Right ahead a dark blur upon the water showed where the *Aurora* lay, and the swirl of white foam behind her spoke of the pace at which she was going. We flashed past barges, steamers, merchant-vessels, in and out, behind this one and round the other. Voices hailed us out of the darkness, but still the *Aurora* thundered on, and still we followed close upon her track.

'Pile it on, men, pile it on!' cried Holmes, looking down into the engine-room, while the fierce glow from below beat upon his eager, aquiline face. 'Get every pound of steam you can.'

'I think we gain a little,' said Jones, with his eyes on the *Aurora*.

'I am sure of it,' said I. 'We shall be up with her in a very few minutes.'

At that moment, however, as our evil fate would have it, a tug with three barges in tow blundered in between us. It was only by putting our helm hard down that we avoided a collision, and before we could round them and recover our way the *Aurora* had gained a good two hundred yards. She was still, however, well in view, and the murky, uncertain twilight was settling into a clear, starlit night. Our boilers were strained to their utmost, and the frail shell vibrated and creaked with the fierce energy which was driving us along. We had shot through the pool, past the West India Docks, down the long Deptford Reach, and up again after rounding the Isle of Dogs. The dull blur in front of us resolved itself now clearly enough into the dainty *Aurora*. Jones turned our searchlight upon her, so that we could plainly see the figures upon her deck. One man sat by the stern, with something black between his knees, over which he stooped.

Beside him lay a dark mass, which looked like a Newfoundland dog.* The boy held the tiller, while against the red glare of the furnace I could see old Smith, stripped to the waist, and shovelling coals for dear life. They may have had some doubt at first as to whether we were really pursuing them, but now as we followed every winding and turning which they took there could no longer be any question about it. At Greenwich we were about three hundred paces behind them. At Blackwall we could not have been more than two hundred and fifty. I have coursed* many creatures in many countries during my chequered career, but never did sport give me such a wild thrill as this mad, flying man-hunt down the Thames. Steadily we drew in upon them, yard by yard. In the silence of the night we could hear the panting and clanking of their machinery. The man in the stern still crouched upon the deck, and his arms were moving as though he were busy, while every now and then he would look up and measure with a glance the distance which still separated us. Nearer we came and nearer. Jones yelled to them to stop. We were not more than four boat's-lengths behind them, both boats flying at a tremendous pace. It was a clear reach of the river, with Barking Level upon one side and the melancholy Plumstead Marshes upon the other. At our hail the man in the stern sprang up from the deck and shook his two clenched fists at us, cursing the while in a high, cracked voice. He was a good-sized, powerful man, and as he stood poising himself with legs astride, I could see that from the thigh downward there was but a wooden stump upon the right side. At the sound of his strident, angry cries, there was movement in the huddled bundle upon the deck. It straightened itself into a little black man—the smallest I have ever seen—with a great, misshapen head and a shock of tangled, dishevelled hair. Holmes had already drawn his revolver, and I whipped out mine at the sight of this savage, distorted creature. He was wrapped in some sort of dark ulster or blanket, which left only his face exposed, but that face was enough to give a man a sleepless night. Never have I seen features so deeply marked with all bestiality and cruelty. His small eyes glowed and burned with a sombre light, and his thick lips were writhed back from his teeth, which grinned and chattered at us with half animal fury.

'Fire if he raises his hand,' said Holmes quietly.

We were within a boat's-length by this time, and almost within touch of our quarry. I can see the two of them now as they stood, the white man with his legs far apart, shrieking out curses, and the unhallowed dwarf with his hideous face, and his strong yellow teeth gnashing at us in the light of our lantern.

It was well that we had so clear a view of him. Even as we looked he plucked out from under his covering a short, round piece of wood, like a school-ruler, and clapped it to his lips. Our pistols rang out together. He whirled round, threw up his arms, and, with a kind of choking cough, fell sideways into the stream. I caught one glimpse of his venomous, menacing eyes amid the white swirl of the waters. At the same moment the wooden-legged man threw himself upon the rudder and put it hard down, so that his boat made straight for the southern bank, while we shot past her stern, only clearing her by a few feet. We were round after her in an instant, but she was already nearly at the bank. It was a wild and desolate place, where the moon glimmered upon a wide expanse of marsh-land, with pools of stagnant water and beds of decaying vegetation. The launch, with a dull thud, ran up upon the mud-bank, with her bow in the air and her stern flush with the water. The fugitive sprang out, but his stump instantly sank its whole length into the sodden soil. In vain he struggled and writhed. Not one step could he possibly take either forward or backwards. He yelled in impotent rage and kicked frantically into the mud with his other foot; but his struggles only bored his wooden pin the deeper into the sticky bank. When we brought our launch alongside he was so firmly anchored that it was only by throwing the end of a rope over his shoulders that we were able to haul him out and to drag him, like some evil fish, over our side. The two Smiths, father and son, sat sullenly in their launch but came aboard meekly enough when commanded. The *Aurora* herself we hauled off and made fast to our stern. A solid iron chest of Indian workmanship stood upon the deck. This, there could be no question, was the same that had contained the ill-omened treasure of the Sholtos. There was no key, but it was of considerable weight, so we transferred it carefully to our own little cabin. As we steamed slowly up-stream again, we flashed our searchlight in every direction, but there was no sign of the Islander. Somewhere in the dark ooze at the bottom of the Thames lie the bones of that strange visitor to our shores.

'See here,' said Holmes, pointing to the wooden hatchway. 'We were hardly quick enough with our pistols.' There, sure enough, just behind where we had been standing, stuck one of those murderous darts which we knew so well. It must have whizzed between us at the instant we fired. Holmes smiled at it and shrugged his shoulders in his easy fashion, but I confess that it turned me sick to think of the horrible death which had passed so close to us that night.

· CHAPTER 11 ·

The Great Agra Treasure

OUR captive sat in the cabin opposite to the iron box which he had done so much and waited so long to gain. He was a sunburned, reckless-eyed fellow, with a network of lines and wrinkles all over his mahogany features, which told of a hard, open-air life. There was a singular prominence about his bearded chin which marked a man who was not to be easily turned from his purpose. His age may have been fifty or thereabouts, for his black, curly hair was thickly shot with gray. His face in repose was not an unpleasing one, though his heavy brows and aggressive chin gave him, as I had lately seen, a terrible expression when moved to anger. He sat now with his handcuffed hands upon his lap, and his head sunk upon his breast, while he looked with his keen, twinkling eyes at the box which had been the cause of his ill-doings. It seemed to me that there was more sorrow than anger in his rigid and contained countenance. Once he looked up at me with a gleam of something like humour in his eyes.

'Well, Jonathan Small,' said Holmes, lighting a cigar, 'I am sorry that it has come to this.'

'And so am I, sir,' he answered frankly. 'I don't believe that I can swing over the job.* I give you my word on the book that I never raised hand against Mr Sholto. It was that little hell-hound, Tonga, who shot one of his cursed darts into him. I had no part in it, sir. I was as grieved as if it had been my blood-relation. I welted the little devil with the slack end of the rope for it, but it was done, and I could not undo it again.'

'Have a cigar,' said Holmes; 'and you had best take a pull out of my flask, for you are very wet. How could you expect so small and weak a man as this black fellow to overpower Mr Sholto and hold him while you were climbing the rope?'

'You seem to know as much about it as if you were there, sir. The truth is that I hoped to find the room clear. I knew the habits of the house pretty well, and it was the time when Mr Sholto usually went down to his supper. I shall make no secret of the business. The best defence that I can make is just the simple truth. Now, if it had been the old major I would have swung for him with a light heart. I would have thought no more of knifing him than of smoking this cigar. But it's cursed hard that I should be lagged* over this young Sholto, with whom I had no quarrel whatever.'

'You are under the charge of Mr Athelney Jones, of Scotland Yard. He is going to bring you up to my rooms, and I shall ask you for a true account of the matter. You must make a clean breast of it, for if you do I hope that I may be of use to you. I think I can prove that the poison acts so quickly that the man was dead before ever you reached the room.'

'That he was, sir. I never got such a turn in my life as when I saw him grinning at me with his head on his shoulder as I climbed through the window. It fairly shook me, sir. I'd have half killed Tonga for it if he had not scrambled off. That was how he came to leave his club, and some of his darts too, as he tells me, which I dare say helped to put you on our track; though how you kept on it is more than I can tell. I don't feel no malice against you for it. But it does seem a queer thing,' he added, with a bitter smile, 'that I, who have a fair claim to half a million of money, should spend the first half of my life building a breakwater in the Andamans, and am like to spend the other half digging drains at Dartmoor.* It was an evil day for me when first I clapped eyes upon the merchant Achmet and had to do with the Agra treasure, which never brought anything but a curse yet upon the man who owned it. To him it brought murder, to Major Sholto it brought fear and guilt, to me it has meant slavery for life.'

At this moment Athelney Jones thrust his broad face and heavy shoulders into the tiny cabin.

'Quite a family party,' he remarked. 'I think I shall have a pull at that flask, Holmes. Well, I think we may all congratulate each other. Pity we didn't take the other alive; but there was no choice. I say, Holmes, you must confess that you cut it rather fine. It was all we could do to overhaul her.'

'All is well that ends well,' said Holmes. 'But I certainly did not know that the *Aurora* was such a clipper.'

'Smith says she is one of the fastest launches on the river, and that if he had had another man to help him with the engines we should never have caught her. He swears he knew nothing of this Norwood business.'

'Neither he did,' cried our prisoner—'not a word. I chose his launch because I heard that she was a flier. We told him nothing; but we paid him well, and he was to get something handsome if we reached our vessel, the *Esmeralda*, at Gravesend, outward bound for the Brazils.'

'Well, if he has done no wrong we shall see that no wrong comes to him. If we are pretty quick in catching our men, we are not so quick in condemning them.' It was amusing to notice how the consequential Jones was already beginning to give himself airs on the strength of the capture. From the slight smile which played over Sherlock Holmes's face, I could see that the speech had not been lost upon him.

'We will be at Vauxhall Bridge presently,' said Jones, 'and shall land you, Dr Watson, with the treasure-box. I need hardly tell you that I am taking a very grave responsibility upon myself in doing this. It is most irregular, but of course an agreement is an agreement. I must, however, as a matter of duty, send an inspector with you, since you have so valuable a charge. You will drive, no doubt?'

'Yes, I shall drive.'

'It is a pity there is no key, that we may make an inventory first. You will have to break it open. Where is the key, my man?'

'At the bottom of the river,' said Small shortly.

'Hum! There was no use your giving this unnecessary trouble. We have had work enough already through you. However, Doctor, I need not warn you to be careful. Bring the box back with you to the Baker Street rooms. You will find us there, on our way to the station.'

They landed me at Vauxhall, with my heavy iron box, and with a bluff, genial inspector as my companion. A quarter of an hour's drive brought us to Mrs Cecil Forrester's. The servant seemed surprised at so late a visitor. Mrs Cecil Forrester was out for the evening, she explained, and likely to be very late. Miss Morstan, however, was in the drawing-room; so to the drawing-room I went, box in hand, leaving the obliging inspector in the cab.

She was seated by the open window, dressed in some sort of white diaphanous material, with a little touch of scarlet at the neck and waist. The soft light of a shaded lamp fell upon her as she leaned back in the basket chair, playing over her sweet grave face, and tinting with a dull, metallic sparkle the rich coils of her luxuriant hair. One white arm and hand drooped over the side of the chair, and her whole pose and figure spoke of an absorbing melancholy. At the sound of my footfall she sprang to her feet, however, and a bright flush of surprise and of pleasure coloured her pale cheeks.

'I heard a cab drive up,' she said. 'I thought that Mrs Forrester had come back very early, but I never dreamed that it might be you. What news have you brought me?'

'I have brought something better than news,' said I, putting down the box upon the table and speaking jovially and boisterously, though my heart was heavy within me. 'I have brought you something which is worth all the news in the world. I have brought you a fortune.'

She glanced at the iron box.

'Is that the treasure then?' she asked, coolly enough.

'Yes, this is the great Agra treasure. Half of it is yours and half is Thaddeus Sholto's. You will have a couple of hundred thousand each. Think of that! An annuity of ten thousand pounds. There will be few richer young ladies in England. Is it not glorious?'

I think I must have been rather over-acting my delight, and that she detected a hollow ring in my congratulations, for I saw her eyebrows rise a little, and she glanced at me curiously.

'If I have it,' said she, 'I owe it to you.'

'No, no,' I answered, 'not to me but to my friend Sherlock Holmes. With all the will in the world, I could never have followed up a clue which has taxed even his analytical genius. As it was, we very nearly lost it at the last moment.'

'Pray sit down and tell me all about it, Dr Watson,' said she.

I narrated briefly what had occurred since I had seen her last. Holmes's new method of search, the discovery of the *Aurora*, the appearance of Athelney Jones, our expedition in the evening, and the wild chase down the Thames. She listened with parted lips and shining eyes to my recital of our adventures. When I spoke of the dart which had so narrowly missed us, she turned so white that I feared that she was about to faint.

'It is nothing,' she said as I hastened to pour her out some water. 'I am all right again. It was a shock to me to hear that I had placed my friends in such horrible peril.'

'That is all over,' I answered. 'It was nothing. I will tell you no more gloomy details. Let us turn to something brighter. There is the treasure. What could be brighter than that? I got leave to bring it with me, thinking that it would interest you to be the first to see it.'

'It would be of the greatest interest to me,' she said. There was no eagerness in her voice, however. It had struck her, doubtless, that it might seem ungracious upon her part to be indifferent to a prize which had cost so much to win.

'What a pretty box!' she said, stooping over it. 'This is Indian work, I suppose?'

'Yes; it is Benares metal-work.'*

'And so heavy!' she exclaimed, trying to raise it. 'The box alone must be of some value. Where is the key?'

'Small threw it into the Thames,' I answered. 'I must borrow Mrs Forrester's poker.'

There was in the front a thick and broad hasp, wrought in the image of a sitting Buddha. Under this I thrust the end of the poker and twisted it outward as a lever. The hasp sprang open with a loud snap. With trembling fingers I flung back the lid. We both stood gazing in astonishment. The box was empty!

No wonder that it was heavy. The ironwork was two-thirds of an inch thick all round. It was massive, well made, and solid, like a chest constructed to carry things of great price, but not one shred or crumb of metal or jewellery lay within it. It was absolutely and completely empty.

'The treasure is lost,' said Miss Morstan calmly.

As I listened to the words and realised what they meant, a great shadow seemed to pass from my soul. I did not know how this Agra treasure had weighed me down, until now that it was finally removed. It was selfish, no doubt, disloyal, wrong, but I could realize nothing save that the golden barrier was gone from between us.

'Thank God!' I ejaculated from my very heart.

She looked at me with a quick, questioning smile.

'Why do you say that?' she asked.

'Because you are within my reach again,' I said, taking her hand. She did not withdraw it. 'Because I love you, Mary, as truly as ever a man loved a woman. Because this treasure, these riches, sealed my lips. Now that they are gone I can tell you how I love you. That is why I said, "Thank God".'

'Then I say "Thank God", too,' she whispered, as I drew her to my side.

Whoever had lost a treasure, I knew that night that I had gained one.

· **CHAPTER 12** ·

The Strange Story of Jonathan Small

A VERY patient man was that inspector in the cab, for it was a weary time before I rejoined him. His face clouded over when I showed him the empty box.

'There goes the reward!' said he gloomily. 'Where there is no money there is no pay. This night's work would have been worth a tenner* each to Sam Brown and me if the treasure had been there.'

'Mr Thaddeus Sholto is a rich man,' I said; 'he will see that you are rewarded, treasure or no.'

The inspector shook his head despondently, however.

'It's a bad job,' he repeated; 'and so Mr Athelney Jones will think.'

His forecast proved to be correct, for the detective looked blank enough when I got to Baker Street and showed him the empty box. They had only just arrived, Holmes, the prisoner, and he, for they had changed their plans so far as to report themselves at a station upon the way. My companion lounged in his armchair with his usual listless expression, while Small sat stolidly opposite to him with his wooden leg cocked over his sound one. As I exhibited the empty box he leaned back in his chair and laughed aloud.

'This is your doing, Small,' said Athelney Jones angrily.

'Yes, I have put it away where you shall never lay hand upon it,' he cried exultantly. 'It is my treasure, and if I can't have the loot I'll take darned good care that no one else does. I tell you that no living man has any right to it, unless it is three men who are in the Andaman convict-barracks and myself. I know now that I cannot have the use of it, and I know that they cannot. I have acted all through for them as much as for myself. It's been the sign of four with us always. Well, I know that they would have had me do just what I have done, and throw the treasure into the Thames rather than let it go to kith or kin of Sholto or Morstan. It was not to make them rich that we did for Achmet. You'll find the treasure where the key is, and where little Tonga is. When I saw that your launch must catch us, I put the loot away in a safe place. There are no rupees* for you this journey.'

'You are deceiving us, Small,' said Athelney Jones sternly; 'if you had wished to throw the treasure into the Thames, it would have been easier for you to have thrown box and all.'

'Easier for me to throw and easier for you to recover,' he answered with a shrewd, side-long look. 'The man that was clever enough to hunt me down is clever enough to pick an iron box from the bottom of a river. Now that they are scattered over five miles or so, it may be a harder job. It went to my heart to do it, though. I was half mad when you came up with us. However, there's no good grieving over it. I've had ups in my life, and I've had downs, but I've learned not to cry over spilled milk.'

'This is a very serious matter, Small,' said the detective. 'If you had helped justice, instead of thwarting it in this way, you would have had a better chance at your trial.'

'Justice!' snarled the ex-convict. 'A pretty justice! Whose loot is this, if it is not ours? Where is the justice that I should give it up to those who have never earned it? Look how I have earned it! Twenty long years in that fever-ridden swamp, all day at work under the mangrove-tree, all night chained up in the filthy convict-huts, bitten by mosquitoes, racked with ague, bullied by every cursed black-faced policeman who loved to take it out of a white man. That was how I earned the Agra treasure, and you talk to me of justice because I cannot bear to feel that I have paid this price only that another may enjoy it! I would rather swing a score of times, or have one of Tonga's darts in my hide, than live in a convict's cell and feel that another man is at his ease in a palace with the money that should be mine.'

Small had dropped his mask of stoicism, and all this came out in a wild whirl of words, while his eyes blazed, and the handcuffs clanked together with the impassioned movement of his hands. I could understand, as I saw the fury and the passion of the man, that it was no groundless or unnatural terror which had possessed Major Sholto when he first learned that the injured convict was upon his track.

'You forget that we know nothing of all this,' said Holmes quietly. 'We have not heard your story, and we cannot tell how far justice may originally have been on your side.'

'Well, sir, you have been very fair-spoken to me, though I can see that I have you to thank that I have these bracelets upon my wrists. Still, I bear no grudge for that. It is all fair and above-board. If you want to hear my story, I have no wish to hold it back. What I say to you is God's truth, every word of it. Thank you, you can put the glass beside me here, and I'll put my lips to it if I am dry.

'I am a Worcestershire man myself, born near Pershore.* I dare say you would find a heap of Smalls living there now if you were to look. I have often thought of taking a look round there, but the truth is that I was never much of a credit to the family, and I doubt if they would be so very glad to see me. They were all steady, chapel-going folk, small farmers, well known and respected over the country-side, while I was always a bit of a rover. At last, however, when I was about eighteen, I gave them no more trouble, for I got into a mess over a girl, and could only get out of it again by taking the Queen's shilling* and joining the 3rd Buffs,* which was just starting for India.

'I wasn't destined to do much soldiering, however. I had just got past the goose-step, and learned to handle my musket, when I was fool enough to go swimming in the Ganges.* Luckily for me, my company sergeant, John Holder,* was in the water at the same time, and he was one of the finest swimmers in the service. A crocodile took me, just as I was half-way across, and nipped off my right leg as clean as a surgeon could have done it, just above the knee. What with the shock and the loss of blood, I fainted, and should have been drowned if Holder had not caught hold of me and paddled for the bank. I was five months in hospital over it, and when at last I was able to limp out of it with this timber-toe strapped to my stump, I found myself invalided out of the Army and unfitted for any active occupation.

'I was, as you can imagine, pretty down on my luck at this time, for I was a useless cripple, though not yet in my twentieth year. However, my misfortune soon proved to be a blessing in disguise. A man named Abel White,* who had come out there as an indigo-planter,* wanted an overseer to look after his coolies and keep them up to their work. He happened to be a friend of our colonel's, who had taken an interest in me since the accident. To make a long story short, the colonel recommended me strongly for the post, and, as the work was mostly to be done on horseback, my leg was no great obstacle, for I had enough knee left to keep a good grip on the saddle. What I had to do was to ride over the plantation, to keep an eye on the men as they worked, and to report the idlers. The pay was fair, I had comfortable quarters, and altogether I was content to spend the remainder of my life in indigo-planting. Mr Abel White was a kind man, and he would often drop into my little shanty and smoke a pipe with me, for white folk out there feel their hearts warm to each other as they never do here at home.

'Well, I was never in luck's way long. Suddenly, without a note of warning, the great mutiny* broke upon us. One month India lay as still and peaceful, to all appearance, as Surrey or Kent; the next there were two hundred thousand black devils let loose, and the country was a perfect hell. Of course you know all about it, gentlemen—a deal more than I do, very like, since reading is not in my line. I only know what I saw with my own eyes. Our plantation was at a place called Muttra,* near the border of the North-west Provinces.* Night after night the whole sky was alight with the burning bungalows, and day after day we had small companies of Europeans passing through our estate with their wives and children, on their way to Agra,* where were the nearest troops. Mr Abel White was an obstinate man. He had it in his head that the affair had been exaggerated, and that it would blow over as suddenly as it had sprung up. There he sat on his veranda, drinking whisky-pegs* and smoking cheroots, while the country was in a blaze about him. Of course we stuck by him, I and Dawson, who, with his wife, used to do the book-work and the managing. Well, one fine day the crash came. I had been away on a distant plantation, and was riding slowly home in the evening, when my eye fell upon something all huddled together at the bottom of a steep nullah.* I rode down to see what it was, and the cold struck through my heart when I found it was Dawson's wife, all cut into ribbons, and half eaten by jackals and native dogs. A little farther up the road Dawson himself was lying on his face, quite dead, with an empty revolver in his hand, and four Sepoys* lying across each other in front of him. I reined up my horse, wondering which way I should turn; but at that moment I saw thick smoke curling up from Abel White's bungalow, and the flames beginning to burst through the roof. I knew then that I could do my employer no good, but would only throw my own life away if I meddled in the matter. From where I stood I could see hundreds of the black fiends, with their red coats still on their backs, dancing and howling round the burning house. Some of them pointed at me, and a couple of bullets sang past my head: so I broke away across the paddy-fields, and found myself at night safe within the walls at Agra.

'As it proved, however, there was no great safety there, either. The whole country was up like a swarm of bees. Wherever the English could collect in little bands they held just the ground that their guns commanded. Everywhere else they were helpless fugitives. It was a fight of the millions against the hundreds; and the cruellest part of it was that these men that we fought against, foot, horse, and gunners, were our own picked troops, whom we had taught and trained, handling our own weapons and blowing our own bugle-calls. At Agra there were the 3rd Bengal Fusiliers,* some Sikhs, two troops of horse, and a battery of artillery. A volunteer corps of clerks and merchants had been formed, and this I joined, wooden leg and all. We went out to meet the rebels at Shahgunge* early in July, and we beat them back, for a time, but our powder gave out, and we had to fall back upon the city.

'Nothing but the worst news came to us from every side—which is not to be wondered at, for if you look at the map you will see that we were right in the heart of it. Lucknow* is rather better than a hundred miles to the east, and Cawnpore* about as far to the south. From every point on the compass there was nothing but torture and murder and outrage.

'The city of Agra is a great place, swarming with fanatics and fierce devil-worshippers of all sorts. Our handful of men were lost among the narrow, winding streets. Our leader moved across the river, therefore, and took up his position in the old fort of Agra. I don't know if any of you gentlemen have ever read or heard anything of that old fort. It is a very queer place—the queerest that ever I was in, and I have been in some rum corners, too. First of all it is enormous in size. I should think that the enclosure must be acres and acres. There is a modern part, which took all our garrison, women, children, stores, and everything else, with plenty of room over. But the modern part is nothing like the size of the old quarter, where nobody goes, and which is given over to the scorpions and the centipedes. It is all full of great deserted halls, and winding passages, and long corridors twisting in and out, so that it is easy enough for folk to get lost in it. For this reason it was seldom that anyone went into it, though now and again a party with torches might go exploring.

'The river washes along the front of the old fort, and so protects it, but on the sides and behind there are many doors, and these had to be guarded, of course, in the old quarter as well as in that which was actually held by our troops. We were short-handed, with hardly men enough to man the angles of the building and to serve the guns. It was impossible for us, therefore, to station a strong guard at every one of the innumerable gates. What we did was to organize a central guard-house in the middle of the fort, and to leave each gate under the charge of one white man and two or three natives. I was selected to take charge during certain hours of the night of a small isolated door upon the south-west side of the building. Two Sikh troopers were placed under my command, and I was instructed if anything went wrong to fire my musket, when I might rely upon help coming at once from the central guard. As the guard was a good two hundred paces away, however, and as the space between was cut up into a labyrinth of passages and corridors, I had great doubts as to whether they could arrive in time to be of any use in case of an actual attack.

'Well, I was pretty proud at having this small command given me, since I was a raw recruit, and a game-legged one at that. For two nights I kept the watch with my Punjaubees.* They were tall, fierce-looking chaps, Mahomet Singh and Abdullah Khan by name, both old fighting men, who had borne arms against us at Chilian Wallah.* They could talk English pretty well, but I could get little out of them. They preferred to stand together and jabber all night in their queer Sikh lingo. For myself, I used to stand outside the gateway, looking down on the broad, winding river and on the twinkling lights of the great city. The beating of drums, the rattle of tomtoms, and the yells and howls of the rebels, drunk with opium and with bang,* were enough to remind us all night of our dangerous neighbours across the stream.

Every two hours the officer of the night used to come round to all the posts, to make sure that all was well.

'The third night of my watch was dark and dirty, with a small driving rain. It was dreary work standing in the gateway hour after hour in such weather. I tried again and again to make my Sikhs talk, but without much success. At two in the morning the rounds passed, and broke for a moment the weariness of the night. Finding that my companions would not be led into conversation, I took out my pipe, and laid down my musket to strike the match. In an instant the two Sikhs were upon me. One of them snatched my firelock up and levelled it at my head, while the other held a great knife to my throat and swore between his teeth that he would plunge it into me if I moved a step.

'My first thought was that these fellows were in league with the rebels, and that this was the beginning of an assault. If our door were in the hands of the Sepoys the place must fall, and the women and children be treated as they were in Cawnpore.* Maybe you gentlemen think that I am just making out a case for myself, but I give you my word that when I thought of that, though I felt the point of the knife at my throat, I opened my mouth with the intention of giving a scream, if it was my last one, which might alarm the main guard. The man who held me seemed to know my thoughts; for, even as I braced myself to it, he whispered: "Don't make a noise. The fort is safe enough. There are no rebel dogs on this side of the river." There was the ring of truth in what he said, and I knew that if I raised my voice I was a dead man. I could read it in the fellow's brown eyes. I waited, therefore, in silence, to see what it was that they wanted from me.

' "Listen to me, Sahib," said the taller and fiercer of the pair, the one whom they called Abdullah Khan. "You must either be with us now, or you must be silenced for ever. The thing is too great a one for us to hesitate. Either you are heart and soul with us on your oath on the cross of the Christians, or your body this night shall be thrown into the ditch, and we shall pass over to our brothers in the rebel army. There is no middle way. Which is it to be—death or life? We can only give you three minutes to decide, for the time is passing, and all must be done before the rounds come again."

' "How can I decide?" said I. "You have not told me what you want of me. But I tell you now that if it is anything against the safety of the fort I will have no truck with it, so you can drive home your knife and welcome."

' "It is nothing against the fort," said he. "We only ask you to do that which your countrymen come to this land for. We ask you to be rich. If you will be one of us this night, we will swear to you upon the naked knife, and by the threefold oath which no Sikh was ever known to break, that you shall have your fair share of the loot. A quarter of the treasure shall be yours. We can say no fairer."

' "But what is the treasure, then?" I asked. "I am as ready to be rich as you can be, if you will but show me how it can be done."

' "You will swear, then," said he, "by the bones of your father, by the honour of your mother, by the cross of your faith, to raise no hand and speak no word against us, either now or afterwards?"

' "I will swear it," I answered, "provided that the fort is not endangered."

' "Then my comrade and I will swear that you shall have a quarter of the treasure which shall be equally divided among the four of us."

' "There are but three," said I.

' "No; Dost Akbar must have his share. We can tell the tale to you while we wait them. Do you stand at the gate, Mahomet Singh, and give notice of their coming. The thing stands thus, Sahib, and I tell it to you because I know that an oath is binding upon a Feringhee,* and that we may trust you. Had you been a lying Hindoo, though you had sworn by all the gods in their false temples, your blood would have been upon the knife and your body in the water. But the Sikh knows the Englishman, and the Englishman knows the Sikh. Hearken, then, to what I have to say.

' "There is a rajah* in the northern provinces who has much wealth, though his lands are small. Much has come to him from his father, and more still he has set by himself, for he is of a low nature, and hoards his gold rather than spend it. When the troubles broke out he would be friends both with the lion and the tiger—with the Sepoy and with the Company's *Raj*.* Soon, however, it seemed to him that the white men's day was come, for through all the land he could hear of nothing but of their death and their overthrow. Yet, being a careful man, he made such plans that, come what might, half at least of his treasure should be left to him. That which was in gold and silver he kept by him in the vaults of his palace, but the most precious stones and the choicest pearls that he had he put in an iron box, and sent it by a trusty servant, who, under the guise of a merchant, should take it to the fort at Agra, there to lie until the land is at peace. Thus, if the rebels won he would have his money, but if the Company conquered, his jewels would be saved to him. Having thus divided his hoard, he threw himself into the cause of the Sepoys, since they were strong upon his borders. By his doing this, mark you, Sahib, his property becomes the due of those who have been true to their salt.

' "This pretended merchant, who travels under the name of Achmet, is now in the city of Agra, and desires to gain his way into the fort. He has with him as travelling-companion my foster-brother Dost Akbar, who knows his secret. Dost Akbar has promised this night to lead him to a side-postern of the fort, and has chosen this one for his purpose. Here he will come presently, and here he will find Mahomet Singh and myself awaiting him. The place is lonely, and none shall know of his coming. The world shall know of the merchant Achmet no more, but the great treasure of the rajah shall be divided among us. What say you to it, Sahib?"

'In Worcestershire the life of a man seems a great and a sacred thing; but it is very different when there is fire and blood all round you, and you have been used to meeting death at every turn. Whether Achmet the merchant lived or died was a thing as light as air to me, but at the talk about the treasure my heart turned to it, and I thought of what I might do in the old country with it, and how my folk would stare when they saw their ne'er-do-weel* coming back with his pockets full of gold moidores.* I had, therefore, already made up my mind. Abdullah Khan, however, thinking that I hesitated, pressed the matter more closely.

'"Consider, Sahib," said he, "that if this man is taken by the commandant he will be hung or shot, and his jewels taken by the Government, so that no man will be a rupee the better for them. Now, since we do the taking of him, why should we not do the rest as well? The jewels will be as well with us as in the Company's coffers. There will be enough to make every one of us rich men and great chiefs. No one can know about the matter, for here we are cut off from all men. What could be better for the purpose? Say again, then, Sahib, whether you are with us, or if we must look upon you as an enemy."

'"I am with you heart and soul," said I.

'"It is well," he answered, handing me back my firelock. "You see that we trust you, for your word, like ours, is not to be broken. We have now only to wait for my brother and the merchant."

'"Does your brother know, then, of what you will do?" I asked.

'"The plan is his. He has devised it. We will go to the gate and share the watch with Mahomet Singh."

'The rain was still falling steadily, for it was just the beginning of the wet season. Brown, heavy clouds were drifting across the sky, and it was hard to see more than a stone-cast. A deep moat lay in front of our door, but the water was in places nearly dried up, and it could easily be crossed. It was strange to me to be standing there with those two wild Punjaubees waiting for the man who was coming to his death.

'Suddenly my eye caught the glint of a shaded lantern at the other side of the moat. It vanished among the mound-heaps, and then appeared again coming slowly in our direction.

' "Here they are!" I exclaimed.

' "You will challenge him, Sahib, as usual," whispered Abdullah. "Give him no cause for fear. Send us in with him, and we shall do the rest while you stay here on guard. Have the lantern ready to uncover, that we may be sure that it is indeed the man."

'The light had flickered onwards, now stopping and now advancing, until I could see two dark figures upon the other side of the moat. I let them scramble down the sloping bank, splash through the mire, and climb half-way up to the gate, before I challenged them.

' "Who goes there?" said I, in a subdued voice.

' "Friends," came the answer. I uncovered my lantern and threw a flood of light upon them. The first was an enormous Sikh, with a black beard which swept nearly down to his cummerbund. Outside of a show I have never seen so tall a man. The other was a little fat, round fellow, with a great yellow turban, and a bundle in his hand, done up in a shawl. He seemed to be all in a quiver with fear, for his hands twitched as if he had the ague, and his head kept turning to left and right with two bright little twinkling eyes, like a mouse when he ventures out from his hole. It gave me the chills to think of killing him, but I thought of the treasure, and my heart set as hard as a flint within me. When he saw my white face he gave a little chirrup of joy and came running up towards me.

' "Your protection, Sahib," he panted, "your protection for the unhappy merchant Achmet. I have travelled across Rajpootana* that I might seek the shelter of the fort at Agra. I have been robbed and beaten and abused because I have been the friend of the Company. It is a blessed night this when I am once more in safety—I and my poor possessions."

' "What have you in the bundle?" I asked.

' "An iron box," he answered, "which contains one or two little family matters which are of no value to others, but which I should be sorry to lose. Yet I am not a beggar; and I shall reward you, young Sahib, and your governor also, if he will give me the shelter I ask."

'I could not trust myself to speak longer with the man. The more I looked at his fat, frightened face, the harder did it seem that we should slay him in cold blood. It was best to get it over.

' "Take him to the main guard," said I. The two Sikhs closed in upon him on each side, and the giant walked behind, while they marched in through the dark gateway. Never was a man so compassed round with death. I remained at the gateway with the lantern.

'I could hear the measured tramp of their footsteps sounding through the lonely corridors. Suddenly it ceased, and I heard voices, and a scuffle, with the sound of blows. A moment later there came, to my horror, a rush of footsteps coming in my direction, with a loud breathing of a running man. I turned my lantern down the long straight passage, and there was the fat man, running like the wind, with a smear of blood across his face, and close at his heels, bounding like a tiger, the great black-bearded Sikh, with a knife flashing in his hand. I have never seen a man run so fast as that little merchant. He was gaining on the Sikh, and I could see that if he once passed me and got to the open air he would save himself yet. My heart softened to him, but again the thought of his treasure turned me hard and bitter. I cast my firelock between his legs as he raced past, and he rolled twice over like a shot rabbit. Ere he could stagger to his feet the Sikh was upon him, and buried his knife twice in his side. The man never uttered moan nor moved muscle, but lay where he had fallen. I think myself that he may have broken his neck with the fall. You see, gentlemen, that I am keeping my promise. I am telling you every word of the business just exactly as it happened, whether it is in my favour or not.'

He stopped and held out his manacled hands for the whisky-and-water which Holmes had brewed for him. For myself, I confess that I had now conceived the utmost horror of the man, not only for this cold-blooded business in which he had been concerned, but even more for the somewhat flippant and careless way in which he narrated it. Whatever punishment was in store for him, I felt that he might expect no sympathy from me. Sherlock Holmes and Jones sat with their hands upon their knees, deeply interested in the story, but with the same disgust written upon their faces. He may have observed it, for there was a touch of defiance in his voice and manner as he proceeded.

'It was all very bad, no doubt,' said he. 'I should like to know how many fellows in my shoes would have refused a share of this loot when they knew that they would have their throats cut for their pains. Besides, it was my life or his when once he was in the fort. If he had got out, the whole business would come to light, and I should have been court-martialled and shot as likely as not; for people were not very lenient at a time like that.'

'Go on with your story,' said Holmes shortly.

'Well, we carried him in, Abdullah, Akbar, and I. A fine weight he was, too, for all that he was so short. Mahomet Singh was left to guard the door. We took him to a place which the Sikhs had already prepared. It was some distance off, where a winding passage leads to a great empty hall, the brick walls of which were all crumbling to pieces. The earth floor had sunk in at one place, making a natural grave, so we left Achmet the merchant there, having first covered him over with loose bricks. This done, we all went back to the treasure.

'It lay where he had dropped it when he was first attacked. The box was the same which now lies open upon your table. A key was hung by a silken cord to that carved handle upon the top. We opened it, and the light of the lantern gleamed upon a collection of gems such as I have read of and thought about when I was a little lad at Pershore. It was blinding to look upon them. When we had feasted our eyes we took them all out and made a list of them. There were one hundred and forty-three diamonds of the first water, including one which has been called, I believe, "the Great Mogul", and is said to be the second largest stone in existence. Then there were ninety-seven very fine emeralds, and one hundred and seventy rubies, some of which, however, were small. There were forty carbuncles, two hundred and ten sapphires, sixty-one agates, and a great quantity of beryls, onyxes, cats'-eyes, turquoises, and other stones, the very names of which I did not know at the time, though I have become more familiar with them since. Besides this, there were nearly three hundred very fine pearls, twelve of which were set in a gold coronet. By the way, these last had been taken out of the chest, and were not there when I recovered it.

'After we had counted our treasures we put them back into the chest and carried them to the gateway to show them to Mahomet Singh. Then we solemnly renewed our oath to stand by each other and be true to our secret. We agreed to conceal our loot in a safe place until the country should be at peace again, and then to divide it equally among ourselves. There was no use dividing it at present, for if gems of such value were found upon us it would cause suspicion, and there was no privacy in the fort nor any place where we could keep them. We carried the box, therefore, into the same hall where we had buried the body, and there, under certain bricks in the best-preserved wall, we made a hollow and put our treasure. We made careful note of the place, and next day I drew four plans, one for each of us, and put the sign of the four of us at the bottom, for we had sworn that we should each always act for all, so that none might take advantage. This is an oath that I can put my hand to my heart and swear that I have never broken.

'Well, there's no use my telling you gentlemen what came of the Indian mutiny. After Wilson* took Delhi and Sir Colin* relieved Lucknow the back of the business was broken. Fresh troops came pouring in, and Nana Sahib* made himself scarce over the frontier. A flying column under Colonel Greathed* came round to Agra and cleared the Pandies* away from it. Peace seemed to be settling upon the country, and we four were beginning to hope that the time was at hand when we might safely go off with our shares of the plunder. In a moment, however, our hopes were shattered by our being arrested as the murderers of Achmet.

'It came about in this way. When the rajah put his jewels into the hands of Achmet he did it because he knew that he was a trusty man. They are suspicious folk in the East, however: so what does this rajah do but take a second even more trusty servant and set him to play the spy upon the first. This second man was ordered never to let Achmet out of his sight, and he followed him like his shadow. He went after him that night and saw him pass through the doorway. Of course he thought he had taken refuge in the fort, and applied for admission there himself next day, but could find no trace of Achmet. This seemed to him so strange that he spoke about it to a sergeant of guides, who brought it to the ears of the commandant. A thorough search was quickly made, and the body was discovered. Thus at the very moment that we thought that all was safe we were all four seized and brought to trial on a charge of murder—three of us because we had held the gate that night, and the fourth because he was known to have been in the company of the murdered man. Not a word about the jewels came out at the trial, for the rajah had been deposed and driven out of India: so no one had any particular interest in them. The murder, however, was clearly made out, and it was certain that we must all have been concerned in it. The three Sikhs got penal servitude for life, and I was condemned to death, though my sentence was afterwards commuted to the same as the others.

'It was rather a queer position that we found ourselves in then. There we were all four tied by the leg and with precious little chance of ever getting out again, while we each held a secret which might have put each of us in a palace if we could only have made use of it. It was enough to make a man eat his heart out to have to stand the kick and the cuff of every petty jack-in-office, to have rice to eat and water to drink, when that gorgeous fortune was ready for him outside, just waiting to be picked up. It might have driven me mad; but I was always a pretty stubborn one, so I just held on and bided my time.

'At last it seemed to me to have come. I was changed from Agra to Madras, and from there to Blair Island* in the

Andamans. There are very few white convicts at this settlement, and, as I had behaved well from the first, I soon found
myself a sort of privileged person. I was given a hut in Hope
Town, which is a small place on the slopes of Mount
Harriet, and I was left pretty much to myself. It is a dreary,
fever-stricken place, and all beyond our little clearings was
infested with wild cannibal natives, who were ready enough
to blow a poisoned dart at us if they saw a chance. There
was digging and ditching and yam-planting, and a dozen
other things to be done, so we were busy enough all day;
though in the evening we had a little time to ourselves.
Among other things, I learned to dispense drugs for the
surgeon, and picked up a smattering of his knowledge. All
the time I was on the lookout for a chance of escape; but it
is hundreds of miles from any other land, and there is little
or no wind in those seas: so it was a terribly difficult job to
get away.

'The surgeon, Dr Somerton, was a fast, sporting young
chap, and the other young officers would meet in his rooms
of an evening and play cards. The surgery, where I used to
make up my drugs, was next to his sitting-room, with a small
window between us. Often, if I felt lonesome, I used to turn
out the lamp in the surgery, and then, standing there, I
could hear their talk and watch their play. I am fond of a
hand at cards myself, and it was almost as good as having
one to watch the others. There was Major Sholto, Captain
Morstan, and Lieutenant Bromley Brown, who were in
command of the native troops, and there was the surgeon
himself, and two or three prison-officials, crafty old hands
who played a nice, sly, safe game. A very snug little party
they used to make.

'Well, there was one thing which very soon struck me, and
that was that the soldiers used always to lose* and the
civilians to win. Mind, I don't say there was anything unfair,
but so it was. These prison-chaps had done little else than
play cards ever since they had been at the Andamans, and
they knew each other's game to a point, while the others just
played to pass the time and threw their cards down anyhow.

Night after night the soldiers got up poorer men, and the poorer they got the more keen they were to play. Major Sholto was the hardest hit. He used to pay in notes and gold at first, but soon it came to notes of hand and for big sums. He sometimes would win for a few deals just to give him heart, and then the luck would set in against him worse than ever. All day he would wander about as black as thunder, and he took to drinking a deal more than was good for him.

'One night he lost even more heavily than usual. I was sitting in my hut when he and Captain Morstan came stumbling along on the way to their quarters. They were bosom friends, those two, and never far apart. The major was raving about his losses.

' "It's all up, Morstan," he was saying as they passed my hut. "I shall have to send in my papers.* I am a ruined man."

' "Nonsense, old chap!" said the other, slapping him upon the shoulder. "I've had a nasty facer myself, but—" That was all I could hear, but it was enough to set me thinking.

'A couple of days later Major Sholto was strolling on the beach: so I took the chance of speaking to him.

' "I wish to have your advice, Major," said I.

' "Well, Small, what is it?" he asked, taking his cheroot from his lips.

' "I wanted to ask you, sir," said I, "who is the proper person to whom hidden treasure should be handed over. I know where half a million worth lies, and, as I cannot use it myself, I thought perhaps the best thing that I could do would be to hand it over to the proper authorities, and then perhaps they would get my sentence shortened for me."

' "Half a million, Small?" he gasped, looking hard at me to see if I was in earnest.

' "Quite that, sir—in jewels and pearls. It lies there ready for anyone. And the queer thing about it is that the real owner is outlawed and cannot hold property, so that it belongs to the first comer."

' "To Government, Small," he stammered, "to Government." But he said it in a halting fashion, and I knew in my heart that I had got him.

' "You think, then, sir, that I should give the information to the Governor-General?"* said I quietly.

' "Well, well, you must not do anything rash, or that you might repent. Let me hear all about it, Small. Give me the facts."

'I told him the whole story, with small changes, so that he could not identify the places. When I had finished he stood stock still and full of thought. I could see by the twitch of his lip that there was a struggle going on within him.

' "This is a very important matter, Small," he said at last. "You must not say a word to anyone about it, and I shall see you again soon."

'Two nights later he and his friend, Captain Morstan, came to my hut in the dead of the night with a lantern.

' "I want you just to let Captain Morstan hear that story from your own lips, Small," said he.

'I repeated it as I had told it before.

' "It rings true, eh?" said he. "It's good enough to act upon?"

'Captain Morstan nodded.

' "Look here, Small," said the major. "We have been talking it over, my friend here and I, and we have come to the conclusion that this secret of yours is hardly a Government matter, after all, but is a private concern of your own, which of course you have the power of disposing of as you think best. Now the question is, What price would you ask for it? We might be inclined to take it up, and at least look into it, if we could agree as to terms." He tried to speak in a cool, careless way, but his eyes were shining with excitement and greed.

' "Why, as to that, gentlemen," I answered, trying also to be cool, but feeling as excited as he did, "there is only one bargain which a man in my position can make. I shall want you to help me to my freedom, and to help my three companions to theirs. We shall then take you into partnership, and give you a fifth share to divide between you."

' "Hum!" said he. "A fifth share! That is not very tempting."

' "It would come to fifty thousand apiece," said I.

' "But how can we gain your freedom? You know very well that you ask an impossibility."

' "Nothing of the sort," I answered. "I have thought it all out to the last detail. The only bar to our escape is that we can get no boat fit for the voyage, and no provisions to last us for so long a time. There are plenty of little yachts and yawls at Calcutta or Madras which would serve our turn well. Do you bring one over. We shall engage to get aboard her by night, and if you will drop us on any part of the Indian coast you will have done your part of the bargain."

' "If there were only one," he said.

' "None or all," I answered. "We have sworn it. The four of us must always act together."

' "You see, Morstan," said he, "Small is a man of his word. He does not flinch from his friends. I think we may very well trust him."

' "It's a dirty business," the other answered. "Yet, as you say, the money will save our commissions handsomely."

' "Well, Small," said the major, "we must, I suppose, try and meet you. We must first, of course, test the truth of your story. Tell me where the box is hid, and I shall get leave of absence and go back to India in the monthly relief-boat to inquire into the affair."

' "Not so fast," said I, growing colder as he got hot. "I must have the consent of my three comrades. I tell you that it is four or none with us."

' "Nonsense!" he broke in. "What have three black fellows to do with our agreement?"

' "Black or blue," said I, "they are in with me, and we all go together."

'Well, the matter ended by a second meeting, at which Mahomet Singh, Abdullah Khan, and Dost Akbar were all present. We talked the matter over again, and at last we came to an arrangement. We were to provide both the officers with charts of the part of the Agra fort, and mark the place in the wall where the treasure was hid. Major Sholto was to go to India to test our story. If he found the box he was to leave it there, to send out a small yacht provisioned for a voyage, which was to lie off Rutland Island,* and to which we were to make our way, and finally to return to his duties. Captain Morstan was then to apply for leave of absence, to meet us at Agra, and there we were to have a final division of the treasure, he taking the major's share as well as his own. All this we sealed by the most solemn oaths that the mind could think or the lips utter. I sat up all night with paper and ink, and by the morning I had the two charts all ready, signed with the sign of four—that is, of Abdullah, Akbar, Mahomet, and myself.

'Well, gentlemen, I weary you with my long story, and I know that my friend Mr Jones is impatient to get me safely stowed in chokey.* I'll make it as short as I can. The villain Sholto went off to India, but he never came back again. Captain Morstan showed me his name among a list of passengers in one of the mail-boats very shortly afterwards. His uncle had died leaving him a fortune, and he had left the Army; yet he could stoop to treat five men as he had treated us. Morstan went over to Agra shortly afterwards, and found, as we expected, that the treasure was indeed gone. The scoundrel had stolen it all without carrying out one of the conditions on which we had sold him the secret. From that time I lived only for vengeance. I thought of it by day and I nursed it by night. It became an overpowering, absorbing passion with me. I cared nothing for the law— nothing for the gallows. To escape, to track down Sholto, to have my hand upon his throat—that was my one thought. Even the Agra treasure had come to be a smaller thing in my mind than the slaying of Sholto.

'Well, I have set my mind on many things in this life, and never one which I did not carry out. But it was weary years before my time came. I have told you that I had picked up something of medicine. One day when Dr Somerton was down with a fever a little Andaman Islander was picked up by a convict-gang in the woods. He was sick to death, and had gone to a lonely place to die. I took him in hand, though he was as venomous as a young snake, and after a couple of months I got him all right and able to walk. He took a kind of fancy to me then, and would hardly go back to his woods, but was always hanging about my hut. I learned a little of his lingo from him, and this made him all the fonder of me.

'Tonga—for that was his name—was a fine boatman, and owned a big, roomy canoe of his own. When I found that he was devoted to me and would do anything to serve me, I saw my chance of escape. I talked it over with him. He was to bring his boat round on a certain night to an old wharf which was never guarded, and there he was to pick me up. I gave him directions to have several gourds of water and a lot of yams, cocoa-nuts, and sweet potatoes.

'He was staunch and true, was little Tonga. No man ever had a more faithful mate. At the night named he had his boat at the wharf. As it chanced, however, there was one of the convict-guard down there—a vile Pathan* who had never missed a chance of insulting and injuring me. I had always vowed vengeance, and now I had my chance. It was as if fate had placed him in my way that I might pay my debt before I left the island. He stood on the bank with his back to me, and his carbine on his shoulder. I looked about for a stone to beat out his brains with, but none could I see.

'Then a queer thought came into my head, and showed me where I could lay my hand on a weapon. I sat down in the darkness and unstrapped my wooden leg. With three long hops I was on him. He put his carbine to his shoulder, but I struck him full, and knocked the whole front of his skull in. You can see the split in the wood now where I hit him. We both went down together, for I could not keep my balance; but when I got up I found him still lying quiet enough. I made for the boat, and in an hour we were well out at sea. Tonga had brought all his earthly possessions with him, his arms and his gods. Among other things, he had a long bamboo spear, and some Andaman cocoa-nut matting, with which I made a sort of a sail. For ten days we were beating about, trusting to luck, and on the eleventh we were picked up by a trader which was going from Singapore to Jiddah with a cargo of Malay pilgrims. They were a rum crowd, and Tonga and I soon managed to settle down among them. They had one very good quality: they let you alone and asked no questions.

'Well, if I were to tell you all the adventures that my little chum and I went through, you would not thank me, for I would have you here until the sun was shining. Here and there we drifted about the world, something always turning up to keep us from London. All the time, however, I never lost sight of my purpose. I would dream of Sholto at night. A hundred times I have killed him in my sleep. At last, however, some three or four years ago, we found ourselves in England. I had no great difficulty in finding where Sholto lived, and I set to work to discover whether he had realized the treasure, or if he still had it. I made friends with someone who could help me—I name no names, for I don't want to get anyone else in a hole—and I soon found that he still had the jewels. Then I tried to get at him in many ways; but he was pretty sly, and had always two prize-fighters, besides his sons and his *khitmutgar*, on guard over him.

'One day, however, I got word that he was dying. I hurried at once to the garden, mad that he should slip out of my clutches like that, and, looking through the window, I saw him lying in his bed, with his sons on each side of him. I'd have come through and taken my chance with the three of them, only even as I looked at him his jaw dropped, and I knew that he was gone. I got into his room that same night, though, and I searched his papers to see if there was any record of where he had hidden our jewels. There was not a line, however, so I came away, bitter and savage as a man could be. Before I left I bethought me that if I ever met my Sikh friends again it would be a satisfaction to know that I had left some mark of our hatred; so I scrawled down the sign of the four of us, as it had been on the chart, and I pinned it on his bosom. It was too much that he should be taken to the grave without some token from the men whom he had robbed and befooled.

'We earned a living at this time by my exhibiting poor Tonga at fairs and other such places as the black cannibal. He would eat raw meat and dance his war-dance: so we always had a hatful of pennies after a day's work. I still heard all the news from Pondicherry Lodge, and for some years there was no news to hear, except that they were hunting for the treasure. At last, however, came what we had waited for so long. The treasure had been found. It was up at the top of the house in Mr Bartholomew Sholto's chemical laboratory. I came at once and had a look at the place, but I could not see how, with my wooden leg, I was to make my way up to it. I learned, however, about a trap-door in the roof, and also about Mr Sholto's supper-hour. It seemed to me that I could manage the thing easily through Tonga. I brought him out with me with a long rope wound round his waist. He could climb like a cat, and he soon made his way through the roof, but, as ill luck would have it, Bartholomew Sholto was still in the room, to his cost. Tonga thought he had done something very clever in killing him, for when I came up by the rope I found him strutting about as proud as a peacock. Very much surprised was he when I made at him with the rope's end and cursed him for a little bloodthirsty imp. I took the treasure-box and let it down, and then slid down myself, having first left the sign of the four upon the table, to show that the jewels had come back at last to those who had most right to them. Tonga then pulled up the rope, closed the window, and made off the way that he had come.

'I don't know that I have anything else to tell you. I had heard a waterman speak of the speed of Smith's launch, the *Aurora*, so I thought she would be a handy craft for our escape. I engaged with old Smith, and was to give him a big sum if he got us safe to our ship. He knew, no doubt, that there was some screw loose, but he was not in our secrets. All this is the truth, and if I tell it to you, gentlemen, it is not to amuse you—for you have not done me a very good turn—but it is because I believe the best defence I can make is just to hold back nothing, but let all the world know how badly I have myself been served by Major Sholto, and how innocent I am of the death of his son.'

'A very remarkable account,' said Sherlock Holmes. 'A fitting wind-up to an extremely interesting case. There is nothing at all new to me in the latter part of your narrative, except that you brought your own rope. That I did not know. By the way, I had hoped that Tonga had lost all his darts; yet he managed to shoot one at us in the boat.'

'He had lost them all, sir, except the one which was in his blow-pipe at the time.'

'Ah, of course,' said Holmes. 'I had not thought of that.'

'Is there any other point which you would like to ask about?' asked the convict affably.

'I think not, thank you,' my companion answered.

'Well, Holmes,' said Athelney Jones, 'you are a man to be humoured, and we all know that you are a connoisseur of crime; but duty is duty, and I have gone rather far in doing what you and your friend asked me. I shall feel more at ease when we have our story-teller here safe under lock and key. The cab still waits, and there are two inspectors downstairs. I am much obliged to you both for your assistance. Of course you will be wanted at the trial. Good-night to you.'

'Good-night, gentlemen both,' said Jonathan Small.

'You first, Small,' remarked the wary Jones as they left the room. 'I'll take particular care that you don't club me with your wooden leg, whatever you may have done to the gentleman at the Andaman Isles.'

'Well, and there is the end of our little drama,' I remarked, after we had sat some time smoking in silence. 'I fear that it may be the last investigation in which I shall have the chance of studying your methods. Miss Morstan has done me the honour to accept me as a husband in prospective.'

He gave a most dismal groan.

'I feared as much,' said he. 'I really cannot congratulate you.'

I was a little hurt.

'Have you any reason to be dissatisfied with my choice?' I asked.

'Not at all. I think she is one of the most charming young ladies I ever met, and might have been most useful in such work as we have been doing. She had a decided genius that way; witness the way in which she preserved that Agra plan from all the other papers of her father. But love is an emotional thing, and whatever is emotional is opposed to that true cold reason which I place above all things. I should never marry myself, lest I bias my judgment.'

'I trust,' said I, laughing, 'that my judgment may survive the ordeal. But you look weary.'

'Yes, the reaction is already upon me. I shall be as limp as a rag for a week.'

'Strange,' said I, 'how terms of what in another man I should call laziness alternate with your fits of splendid energy and vigour.'

'Yes,' he answered, 'there are in me the makings of a very fine loafer, and also of a pretty spry sort of fellow. I often think of those lines of old Goethe:

Schade dass die Natur nur *einen* Mensch aus dir schuf,
Denn zum würdigen Mann war und zum Schelmen der Stoff.*

By the way, *à propos* of this Norwood business, you see that they had, as I surmised, a confederate in the house, who could be none other than Lal Rao, the butler: so Jones actually has the undivided honour of having caught one fish in his great haul.'

'The division seems rather unfair,' I remarked. 'You have done all the work in this business. I get a wife out of it, Jones gets the credit, pray what remains for you?'

'For me,' said Sherlock Holmes, 'there still remains the cocaine-bottle.' And he stretched his long white hand up for it.

EXPLANATORY NOTES

Lippincott's Magazine of Philadelphia first published the story in its Feb. 1890 issue under the title 'The Sign of the Four; or, The Problem of the Sholtos'.

ACD's pocket diary for 1889 notes on 30 Sept.: '"The Sign of the Four" finished & dispatched', although the shortened form of the title appears in later entries. The manuscript of the story confuses the situation in respect of the story's title still further. It appears that there may have been a facing sheet in ACD's hand on which he had written 'The Sign of Four'. The first page, however, is headed (in a hand other than ACD's) 'The Sign of the Four'. The author used both forms of the title, although he may have found the shorter form more convenient. From the point of view of published editions, the title used depended on whether the story was set from the magazine or from an authorized edition. Pirated editions tended to use the longer title, although this is not always the case.

The first British edition of the book was published as *The Sign of Four* by Spencer Blackett in Oct. 1890. Throughout the story ACD refers to 'the sign of the four'; since this is consistent with the form of the title originally intended for publication, *The Sign of the Four* has been adopted as the title for this edition.

3   *Sherlock Holmes*: the original manuscript of *The Sign of the Four* survives, in private keeping. The first folio sheet was at some point detached from the remainder and was auctioned in New York on 22 Nov. 1929, realizing $50; it has since changed hands several times for much larger sums. Baring-Gould reproduced it (I. 689) and it suggests strongly that unless there was a preliminary draft—unlikely but not impossible at this early state—the beginning at least gave its author little trouble once his mind had settled on the story and its strategy. The text ran from 'Sherlock Holmes took his bottle . . .' to 'His great powers, his masterly manner, and the experience which I had had of his many.' *Corrections in MS*:

'left' is inserted before 'shirt-cuff'.

'sinewy' is substituted for 'thin' before 'forearm'.

'tiny' substituted immediately before 'piston' for illegible word, probably beginning 't' ending 'y', conceivably 'trusty'.

'Swelled' substituted for 'swoll', suggesting the sentence was originally given 'swollen' as past participle governed by 'had' in tandem with preceding 'become'; but the author saw the prolixity—and poor syntax—into which he was falling and rescued the sentence by 'swelled' before writing 'nightly'.

hypodermic syringe: the American surgeon William Stewart Halsted (1852–1922) discovered in 1884 that a whole region of the body could be anaesthetized by injecting cocaine into the nerve. There may be an implication that Holmes had begun such practices as auto-experimentation, only to become addicted to the drug. Something of this kind happened to Halsted and to the popularizers of chloroform for midwifery in ACD's native Edinburgh.

Beaune: fine Burgundian wines produced in the countryside around Beaune in France.

morphine or cocaine: this is the only reference to Holme's addiction to morphine. As late as 1887 cocaine, a white crystalline drug prepared from the leaves of the South American coca plant, was considered in Europe to be a wonder drug. There would have been little public reaction to Holmes's use of the drug as it was easily obtainable in those days. Cocaine was used in snuff, sweets, ointments, and Coca Cola until the early years of the twentieth century. Rodin and Key (*The Medical Casebook of Doctor Arthur Conan Doyle*) observe, 'Cocaine users do not develop a physical basis for their craving. But there is a very strong psychological craving which also constitutes an addiction.'

black-letter volume: a book printed in black letter, the Gothic or Old English style of lettering, which was introduced into England in the mid-fourteenth century and was the character generally used in early printed books.

4   *the Afghan campaign*: in 1878 Sher Ali (1825–79), who had become Emir in 1863, was invited to receive a British mission sent as a consequence of Russian diplomatic overtures. His refusal precipitated the Second Afghan War of 1878–9, during the course of which Sher Ali died. His son Yakub Khan (1849–1923) abdicated in favour of Sher Ali's nephew Abdur Rahman. In July 1880 Sher Ali's younger son, Ayub Khan

(1855–1914), led an uprising against his cousin's acceptance of the British peace terms. Ayub Khan marched upon Kandahar and defeated a British force at Maiwand. He was finally defeated in September 1881, when Abdur Rahman's rule was confirmed. ACD chose the Battle of Maiwand as the battle in which Watson received the wound that was to invalid him out of the army (see *A Study in Scarlet*). The extent of the British defeat at Maiwand should not be overlooked and the following account of the battle is given by Field Marshal Frederick Roberts, first Earl Roberts (1832–1914) in his memoirs *Forty-one Years in India* (1911):

'On the afternoon of the 26th [July 1880] information was received by Brigadier General Burrows that 2,000 of the enemy's Cavalry and a large body of *ghazis* had arrived at Maiwand, eleven miles off, and that Ayub Khan was about to follow with the main body of his army.

'To prevent Ayub Khan getting to Ghazni, General Burrows had to do one of two things, either await him at Khushk-i-Nakhud, or intercept him at Maiwand . . . he determined to adopt the latter course, as he hoped thus to be able to deal with the *ghazis* before they were joined by Ayub Khan.

'. . . At 10 a.m., when about half-way to Maiwand, a spy brought in information that Ayub Khan had arrived at that place, and was occupying it in force; General Burrows, however, considered it then too late to turn back, and decided to advance. At a quarter to twelve the forces came into collision, and the fight lasted until past three o'clock. The Afghans, who, Burrows reported, numbered 25,000, soon outflanked the British. Our Artillery expended their ammunition, and the Native portion of the brigade got out of hand, and pressed back on the few British Infantry, who were unable to hold their own against the overwhelming numbers of the enemy. Our troops were completely routed, and had to thank the apathy of the Afghans in not following them up for escaping total annihilation.

'Of the 2,476 men engaged at Maiwand, 934 were killed and 175 were wounded and missing; the remnant struggled on throughout the night to Kandahar.'

4 *But consider . . . cost!*: ACD was aware of the detrimental effects of cocaine well in advance of general recognition of the fact, which did not occur until the early 1900s.

the game is hardly worth the candle: from seventeenth-century sentiment that the cost of illumination is greater than the benefit from foolish games.

Gregson or Lestrade or Athelney Jones: Detective Inspectors Tobias Gregson and G. Lestrade of the Criminal Investigation Department of the London Metropolitan Police at Scotland Yard appear in *A Study in Scarlet*. The reader has not yet had the privilege of meeting their colleague Athelney Jones, whose name may have been inspired by that of the Radical MP for Durham NW (1885–1914), Llewellyn Archer Atherley-Jones, KC (1851–1929). Athelney Jones in the *Sign*, read aloud, works well in a Welsh voice and may have been conceived as such.

5 *A Study in Scarlet*: for a comment on the ensuing passage see introduction to the companion volume in this series, *A Study in Scarlet*.

the fifth proposition of Euclid: Euclid (*c.*300 BC) was a Greek (probably Alexandrian) mathematician famous for his *Elements*, in which he derived all that was known of geometry from a few simple axioms, definitions, and postulates. Proposition 5 of the first book is: 'if two sides of a triangle are equal, then the angles opposite these sides are equal: should the sides be extended below the base of the triangle, the angles thus formed are also equal to one another.'

Jezail bullet: the Jezail was a heavy, long-barrelled musket manufactured and used by Asians.

François le Villard: a name chosen to resemble François Villon (*b.* 1431), French poet and criminal.

6 *'magnifiques', 'coup-de-mâitres', and 'tours de force'*: 'magnificent', 'master-strokes', and 'feats of strength or skill'.

lunkah: a thin cigar, open at both ends. It was Indian in manufacture and resembled a cheroot.

Trichinopoly: a cigar made of the dark tobacco manufactured near Trichinopoly in southern India.

bird's-eye: a style of cut tobacco.

7 *Wigmore Street Office*: 'Seymour Street Office' in *Lippincott's*. ('Wigmore Street Post-Office' as here and in first English editions, for first mention.) The streets are linked by Portman Square south, Wigmore Street thus intersecting with Baker

Street from the east. ACD, writing in Southsea, did not yet know of the change of name to 'Wigmore Street' from 'Upper Seymour Street', in which the post office was located, and, learning of it, made one change but forgot the other. He rectified this for the book text in his letter to Stoddart of 6 Mar. 1890.

8 *Eliminate all other factors*: one of Holmes's more famous aphorisms, which he uses in at least three other investigations, 'The Beryl Coronet' (*Adventures*), 'The Blanched Soldier' (*Case-Book*), and 'The Bruce-Partington Plans' (*His Last Bow*).

10 *fifty-guinea*: £52.50. The guinea disappeared with the decimalization of Britain's coinage in 1971. The guinea coin was last minted in Britain in 1813 but, even as late as 1970, many shops still quoted prices in guineas. Many professional fees were also quoted in this way.

11 *Morstan*: the first name to incorporate the most common of all first syllables of surnames in the Holmes cycle (Moran, Morcar, Morecroft, Morgan, Moriarty, Morland, Morphy, Morris, Morrison, Mortimer, Morton; and the ultimately unused Mordhouse intended for 'The Lion's Mane' (*Case-Book*)). Evidently from 'mors' (Latin), death. Atmospheric in effect, presumably intentional.

13 *Langham Hotel*: in Langham Place, off the north terminus of Regent Street. One of the most prestigious of London's hotels, which would have been in ACD's mind as the location of Stoddart's repast at which *The Sign of the Four* was commissioned. It subsequently became part of the British Broadcasting Corporation, as additional studios and offices to the main BBC radio premises opposite, in Portland Place. The building reopened as a luxury hotel in 1991.

the Andaman Islands: a chain of islands on the eastern side of the Bay of Bengal, 204 in number, 120 miles from Cape Negrais in Burma, and 340 miles from the northern extremity of Sumatra. Opened as a convict settlement in 1858, during the Great Indian Mutiny.

14 *September 7*: 'By the way there is one very obvious mistake which must be corrected in book form—in the second chapter the letter is headed July 7th, and on almost the same page I talk of its being a September evening' (ACD to J. M. Stoddart,

6 Mar. 1890: *Uncollected Sherlock Holmes*, 50). It does not seem to have been corrected until now.

sixpence: 2½ new pence. The sixpence was first minted in 1550 and disappeared from the British coinage following decimalization in 1971.

Lyceum Theatre: built by Dr Samuel Arnold in 1794 as an opera-house. From 1878 to 1902 it was managed by (Sir) Henry Irving, before it became a music hall. Irving seems to have been one of the many origins for Holmes; ACD had venerated him as a performer from his youth. The Lyceum was in Wellington Street, which crosses the Strand.

16 *What do you make of this fellow's scribble?*: the use of graphology as a plot device is a recurring theme in the Holmes stories, but its importance as a means of deduction is given most emphasis in 'The Reigate Squire' (*Memoirs*). In the nineteenth century Abbé Flandrin, Abbé Michon, and Desbarolles pioneered the analytical system of graphology in France by establishing a technique of fixed signs. In the early 1880s Crepieux-Jamin observed that handwriting is gesture, in a circumscribed, subtle, and infinitely variable way, and that these gestures can express changing moods, emotion, and bodily well-being. Crepieux-Jamin's theory was scientifically tested by others, including Alfred Binet (1857–1911), who pioneered the principles used in intelligence tests. Systems of graphology developed from it and the science is now commonly used, particularly by personnel assessors. ACD was to draw on graphology to support his arguments to prove the innocence of George Edalji during his investigation of that case in 1907.

Winwood Reade's 'Martyrdom of Man': William Winwood Reade (1838–75) was an explorer, novelist, and nephew of Charles Reade (1814–84) whose *Cloister and the Hearth* ACD, in his *Through the Magic Door* (1907), classed as 'our greatest historical novel, and indeed, as being our greatest novel of any sort'. *The Martyrdom of Man* exposes the author's inclination to atheism and was also an influence on H. G. Wells (1866–1946) who described it as 'an extraordinarily inspiring presentation of human history as one consistent process'.

pathology: the branch of medicine concerned with the study of disease and disease processes in order to understand their causes and nature. The specialty originated in the mid-nineteenth century, when Rudolf Virchow (1821–1902) demonstrated that changes in the structure of cells and tissues were related to specific diseases. Today pathology includes studies of the chemistry of blood, urine, and diseased tissue, obtained by biopsy or at autopsy. Watson's choice of reading matter indicated an intention of returning to his profession, which his engagement and marriage later hardened. He is still an invalid on pension in this book.

17 *back files of 'The Times'*: Samuel Palmer published a quarterly index to *The Times*.

18 *four-wheeler*: a four-wheeled enclosed cab drawn by a single horse. Officially known as a clarence, unofficially as a growler.

19 *It was a September evening*: the story was being written in September and the young author forgot that the letter to Miss Morstan was postmarked '7 July' (see previous note, p. 14).

hansoms: the hansom was a two-wheeled covered carriage drawn by a single horse, named after its patentee Joseph Aloysius Hansom (1803–82). It held two people in addition to the driver, who was mounted on an elevated seat behind the body of the carriage.

20 *street arab*: a neglected or abandoned boy or girl of the streets.

21 *Robert Street*: since 1872 part of Stockwell Place, and from 1890 part of Robsart Street. ACD, writing in Southsea, was using an old map, possibly one belonging to his father.

Hindoo (or Hindu): one of the people speaking the Hindi dialect of the North-West Provinces; one of the Aryan inhabitants of northern India, or one who professes the Hindu religion.

Sahib: a term of respect used by Indians.

khitmutgar: Hindi for butler or manservant.

22 *from among it like a mountain-peak from fir-trees*: this description of Thaddeus Sholto has a parallel in the description of the decadent aesthete Lionel Dacre in ACD's short story 'The Leather Funnel' (1903): 'his huge, domelike skull, which curved upwards from amongst his thinning locks, like a snow-peak above its fringe of fir trees'. But Dacre is much more formidable than Thaddeus Sholto, whose upper head suggests the poet Swinburne.

Nature had given him a pendulous lip, and a too visible line of yellow and irregular teeth: Lionel E. Fredman suggests: 'There is something unmistakably Wildean about Thaddeus Sholto. Wilde was an incessant and affected smoker; so was Thaddeus—at least with his hookah. Both were constantly passing their hands over the lower part of the face, self-conscious of discoloured teeth but, in fact, drawing attention to them' (*ACD—The Journal of the Arthur Conan Doyle Society*, 1, 2, 91–3). Sholto's aesthetic prejudices are closer to popular caricatures of Wilde than to Wilde himself. He was also more of a homebody. His handwriting has some points in common with that of Wilde, from whom ACD had correspondence, probably beginning after their meeting with Stoddart.

hookah: a tobacco-pipe of Eastern origin in which the smoke is inhaled through a long flexible tube which draws it through a bowl of perfumed water.

23 *mitral valve*: the valve in the heart that guards the opening between the left auricle and the left ventricle, and prevents the blood in the ventricle from returning to the auricle.

aortic: the valve in the heart that separates the left ventricle from the aorta, which carries blood from the heart to practically all parts of the body.

alive now: tactless, but tactical in its preparation of ground to establish Morstan's death as accidental—which it may not have been.

24 *Chianti*: a wine produced in the region of hills in north-central Italy, which lie between Florence and Siena.

Tokay: a heavy, rich, sweet wine produced in northern Hungary.

Corot: Jean Baptiste Camille Corot (1796–1875). A French landscape painter who visited Italy in 1825–8, making it the subject of many of his landscapes. He exhibited at the Paris *salon* from 1827, but did not achieve critical acclaim until the 1850s, with his misty landscapes populated by nymphs. Today, he is more popularly regarded for his open-air sketches, small landscapes, and figure studies.

though a connoisseur . . . Salvator Rosa: Salvator Rosa (1615–73) was an Italian painter and etcher whose wild and romantic landscapes were much admired in eighteenth-century England. In referring to a doubtful Salvator Rosa, ACD recalls two land-scape paintings attributed to Salvator Rosa at the Jesuit Stonyhurst College in Lancashire, which he attended. The paintings were included in the College's catalogue of 1866 as in the 'Style of Salvator Rosa'. The Revd F. J. Turner, SJ, Archivist to Stonyhurst College, and a former headmaster, writes: 'At some time they were at St Mary's Hall, which was the house of study for young Jesuit students for the priesthood, about two hundred yards from the College, but they may have been removed there later. They are still in the College here. I think now very few would believe that they are really by Salvator Rosa.'

24 *Bouguereau*: Adolphe William Bouguereau (1825–1905) was a French academic and painter of religious and mythological studies.

the modern French school: pompous way of saying 'recent French painting'.

25 *Pondicherry*: a town on the east coast of India.

prize-fighters: bare-knuckle fighters for money, an activity out-lawed since 1867, when the Queensberry rules were drawn up under the direction of John Sholto Douglas, eighth Marquess of Queensberry (1844–1900). The use of his middle name as a surname in *The Sign of the Four* may have arisen from this. That both Queensberry's name and aspects of Wilde appeared in Thaddeus was an extraordinarily prophetic coincidence: the two had not yet met, but Queensberry was to ruin Wilde in 1895.

26 *chaplet*: a small coronet.

30 *Le mauvais goût mène au crime*: 'bad taste leads to crime.' A phrase immortalized by the French novelist Henri Beyle ('Stendhal') (1783–1842).

Astrakhan: dark, curly fleece of very fine wool, Persian or Syrian in origin, derives its name from a city in European Russia, where the wool of the very young lambs closely resembling fur is used. Wilde wore it. So did Mr Gladstone.

valetudinarian: a person of delicate health, and fussy about it.

31 *so he worked out all the cubic space of the house*: Holmes was to use a similar method during the course of his deductions in 'The Norwood Builder' (*Return*).

33 *that amateur*: American editions' usage. English editions, save first and Penguin, have 'the amateur' but 'that amateur' is characteristic early Conan Doyle (e.g. story 'That Veteran').

benefit: a sporting event, the proceeds of which would have been given to McMurdo.

fancy: (slang) prize-fighters. A little archaic by 1888, perhaps.

35 *a hill near Ballarat*: a city and gold-mining area in the state of Victoria, Australia, which was the centre of one of the richest gold-yielding regions in the world.

40 *snibbed*: fastened.

41 *Senegambia*: a region of West Africa whose name is taken from the rivers Senegal and Gambia which lie within its borders.

How often . . . truth?: the fuller version of one of Holmes's more famous aphorisms (see note to page 8).

43 *shire*: Holmes is probably using the term loosely to denote a county.

the rule of three: an arithmetical principle which stated that 'if three quantities of a proportion are known, the fourth can be determined, since the product of the means equals the products of the extremes'.

rigor mortis: stiffening of the body caused by the contraction of the muscles after death.

Hippocratic smile, or 'risus sardonicus': a spasmodic grin exhibited during tetanus, named after Hippocrates (?460–?360 BC), who made the earliest recorded notes of many symptoms of disease and death.

tetanus: a serious disease caused by the bacterium *clostridium tetani* entering wounds and producing a powerful toxin which irritates the nerves supplying muscles.

46 *Il n'y a pas des sots si incommodes que ceux qui ont de l'esprit*: the usual translation is 'There are no fools so troublesome as those who have some wit.' From *Les Maximes* (no. 451) of François, duc de la Rochefoucauld (1613–80). 'I couldn't find this one anywhere and finally concluded that Holmes has simply paraphrased Molière's line about an erudite fool being more foolish than an ignorant one' (Eric Ambler, introduction to *The Adventures of Sherlock Holmes* [1974]).

46 *anything which you may say will be used against you*: exceptionally savage parody of the usual warning on arrest 'anything you say will be taken down and may be used in evidence' (to which 'against you' is frequently but erroneously added).

47 *mare's-nest*: an illusory but much-trumpeted discovery.

Toby: it is interesting that ACD should have chosen the name Toby for the dog in the story. Toby was the trained dog introduced into the Punch and Judy show in the first half of the nineteenth century, and was included in the sketch made by ACD's uncle, Richard Doyle (1824–83), to illustrate *Punch* magazine's best-known cover.

48 *Wir sind gewohnt dass die Menschen verhöhnen was sie nicht verstehen*: 'We are used to seeing that Man despises what he does not understand.' The quotation is from Goethe's *Faust*, part 1.

Goethe: Johann Wolfgang von Goethe (1749–1832), German poet, dramatist, novelist, and scientist.

49 *a half-pay surgeon*: a surgeon drawing his [army] pension.

50 *So help me gracious*: i.e. 'gracious God'; or 'So help me by the grace of God'. Forensic rather than religious in origin.

wiper: viper. A name generally applied to snakes having only two poisonous fangs in the upper jaw. Cockneys in the nineteenth century were popularly supposed to pronounce 'v' as 'w' (see Charles Dickens's works).

slowworm: a harmless, snake-like lizard.

51 *guyed at*: made fun of, sneered at.

lurcher: a breed of hunting-dog.

52 *bull's-eye*: a lantern with a focusing lens.

cord: corrected from 'card', a misreading of ACD's handwriting accidentally perpetuated in all subsequent editions.

Blondin: Charles Blondin (1824–97) an alias of Jean François Gravelet, who crossed the Niagara Falls on a tightrope in 1855.

53 *pocket*: possibly 'packet'.

Martini bullet: a bullet from the Martini–Henry rifle used by British forces.

56 *Mohammedans*: followers of Muhammad (?570–632) who, according to Muslims, was the last of the prophets and preacher

of Islam to the Arabs. After his death, Muhammad's revelations were collected to form the Koran.

57 *tendo Achillis*: the tendon of the heel, so named because it was the only vulnerable part of the body of Achilles who, according to Greek mythology, was the greatest Greek warrior in the Trojan war.

58 *Jean Paul*: popular form of allusion to Johann Paul Friedrich Richter (1763–1825), a German author best known for romances, humorous works, and philosophical treatises.

I worked back to him through Carlyle: Thomas Carlyle (1795–1881), Scottish historian and essayist, worked unhappily as a teacher until 1819. He moved to London in 1834 and published *Sartor Resartus*, a blend of fiction, philosophy, and autobiography, in 1836. This was followed in 1837 by his major work, *The French Revolution*, which, along with later works, expresses his view of history as shaped by the inspired individual. He popularized many German writers in Britain, notably Richter. He retired from public life, grief-stricken at the death of his wife, in 1866.

61 *shillin'*: a shilling, the equivalent of 5 new pence.

63 *wherry*: a light, shallow rowing-boat.

sheets of the wherry: the space at the bow or stern of the boat.

64 *wharfingers*: wharf-owners.

Millbank Penitentiary: the huge penitentiary at Millbank, built in 1813, was designed on lines suggested by Jeremy Bentham's Panopticon. It was comparatively sanitary, but life on the inside was monotonous and lonely. Only during the second half of their sentences were prisoners allowed to leave their separate cells to work with fellow-inmates. The penitentiary was demolished in 1890 to make way for the Tate Gallery which was built on its site.

66 *three bob and a tanner*: 3 shillings and 6 pence (17½ new pence).

68 *Port Blair*: the chief settlement of the Andaman Islands, first occupied in 1789 and refounded in 1856.

The aborigines of the Andaman Islands . . . gained: ACD is quoting the popular view of the inhabitants of the islands, probably from the collection of early Arab notes on India and China (AD 851): 'The inhabitants of these islands eat men alive. They are black with woolly hair, and in their eyes and countenances there is something quite frightful . . . They go naked, and have no boats. If they had, they would devour all who passed near them.' The ninth edition of the *Encyclopaedia Britannica* (1890) stated: 'The Andaman countenance has generally impressed Europeans at first as highly repulsive . . . The people, especially the men, are often robust and vigorous, though their stature is low—seldom five feet, and generally much less.' The eleventh edition (1910) declared: 'The figures of the men are muscular and well-formed and generally pleasing . . . the young men are often distinctly good-looking'. It denied cannibalism, doubted if it had ever been true, but admitted massacres of shipwrecked crews: 'a people to like but not to trust'. In fact they were likeable, and too trusting.

70 *half-sovereign*: a British gold coin, now rarely used. Originally worth a nominal 10 shillings (50 new pence). Modern-day sovereigns and half-sovereigns have a value linked to the current price of gold.

71 *cooling medicine*: medicine believed capable of lowering the temperature of the blood.

73 *pea-jacket*: a heavy coat originally worn by seamen.

75 *outré*: eccentric.

bandanna: coloured handkerchief with yellow or white spots.

76 *a most promising officer*: this well-meant if heavy-footed attempt at a compliment seems an ironic use of Sergeant Cuff on Gooseberry (see introduction).

78 *I thought my disguise . . . test*: 'You would have made an actor and a rare one', remarks Athelney Jones and we may agree, judging by the number of disguises in which Holmes appears. Disguise frequently enabled him to infiltrate where he would not otherwise have been able to. Particularly fine examples are found in 'A Scandal in Bohemia', 'The Man with the Twisted Lip', 'The Beryl Coronet' (*Adventures*), 'The Final Problem' (*Memoirs*), 'The Dying Detective', and 'His Last Bow' (*His Last Bow*). ACD himself enjoyed dressing up in disguise and, following the completion of *The Lost World*, posed as Professor Challenger for the expedition photograph that appeared in the book's first edition. Disguised as Challenger, he also called on his brother-in-law, Ernest William Hornung (1866–1921), the

creator of Raffles, whom he proceeded to deceive with his disguise.

Westminster Stairs: more likely Westminster Pier which is located on the Victoria Embankment just below Westminster Bridge.

79 *miracle plays*: medieval productions of biblical or hagiographical subject. The four great English collections are performed at York, Chester, Coventry, and Wakefield. Also known as 'mysteries'; acted by guildsmen on wheeled stages; principally played on great Festivals. ACD attended a Portsmouth Literary and Scientific Society meeting on 'Miracle Plays' on 15 Feb. 1887.

80 *bon vivant*: a lover of good living.

bumper: a glass of wine filled to the brim, usually so styled in association with a toast, or 'at parting'. Hence Sir Toby Bumper, in Sheridan's *The School for Scandal*.

the Tower: the Tower of London.

81 *One of our greatest statesmen . . . rest*: an aphorism usually attributed to William Ewart Gladstone (1809–98), by then thrice Prime Minister of Great Britain, with a further premiership to come. Its occasion is uncertain (though he probably said it more than once), but it is associated with press interviews while he was cutting trees. Presumably Holmes is not alluding to Karl Marx (1818–83): 'Constant labour of one uniform kind destroys the intensity and flow of a man's animal spirits, which find recreation and delight in mere change of activity' (*Das Kapital*, Part II, Chapter 9).

82 *the Downs*: a sheltered anchorage for shipping between the coast of Kent and the Goodwin Sands.

84 *'a priori'*: presumptive.

86 *Newfoundland dog*: a large black hunting-dog with a bushy coat.

coursed: pursued.

89 *I can swing over the job*: be hanged for the job.

lagged: (slang) arrested.

90 *digging drains at Dartmoor*: a popular expression to describe Princetown Prison on Dartmoor, originally built in 1809 to house French prisoners of war. It was rebuilt and adapted for its present use in 1850.

93 *Benares metal-work*: Benares is an Indian city located on the Ganges. It is the headquarters and most holy city of the Hindu religion and is famous for textiles, jewellery, and brasswork.

94 *tenner*: £10.

95 *rupees*: silver coins of British India, thus prompting Wilde: '*Miss Prism*: . . . The chapter on the Fall of the Rupee you may omit. It is somewhat too sensational. Even these metallic problems have their melodramatic side' (*The Importance of Being Earnest*, Act II).

96 *Pershore*: a market town in south Worcestershire, on the edge of the Vale of Evesham.

97 *taking the Queen's shilling*: to join the Army. The expression derives from the old custom of British Army recruiting-officers paying a shilling to each new recruit.

the 3rd Buffs: more properly, the 3rd East Kent Regiment of Foot, an infantry regiment of the British Army dating from 1665.

Ganges: the most important river in India, both in size and in its religious significance to the Hindus.

Holder: so printed in *Lippincott*'s. The first British book edition had 'Holders', but 'Holder' elsewhere in the same chapter. Later British editions, including Author's edition, have 'Holder'.

Abel White: the original *Lippincott*'s text had this as one word, 'Ablewhite'. The name is reminiscent of the 'Maple White' in *The Lost World*, and to the name 'Godfrey Ablewhite' in *The Moonstone*.

indigo-planter: indigo is a blue vegetable dye extracted from a plant which grows in India.

the great mutiny: The Indian Mutiny (1857–9) was a revolt of some 35,000 sepoys (Indian soldiers in the service of the British East India Company), which developed into a bloody Anglo-Indian war. British reinforcements under their commander-in-chief, Sir Colin Campbell (1792–1863), regained Delhi and relieved Lucknow in late 1857. By July 1858 the revolt had been more or less contained.

98 *Muttra*: an Indian city some thirty miles from Agra.

North-west Provinces: a political division of British India. It was created in 1835, enlarged in 1877 and renamed the United Provinces in 1902.

Agra: a city in north-central India, located on the River Ganges. It is famous for the Taj Mahal.

whisky-pegs: slang for a highball (whisky or brandy with soda). The usual explanation is that each drink represents a peg in the coffin of the drinker.

nullah: Hindi word for ravine or valley.

Sepoys: a native British Indian soldier.

99 *3rd Bengal Fusiliers*: probably the Third Bengal Infantry which remained loyal during the Mutiny.

Shahgunge: a western suburb of Agra.

Lucknow: an Indian city that was besieged during the Mutiny.

Cawnpore: (more correctly, Kanpur) an Indian city situated on the Ganges.

100 *Punjaubees*: (correctly, Punjabis) the Punjab is a region of India and Pakistan below the Himalayan foothills on the flat alluvial plain of five tributaries of the River Indus. Sixty per cent of the population living in India's Punjab state today is Sikh; Pakistan's Punjabi population is almost entirely Muslim.

Chilian Wallah: a battle fought on 13 Jan. 1849 in which the British won the territory of the Punjab.

bang: (or bhang) Indian hemp, smoked or chewed as a narcotic.

101 *the women and children be treated as they were in Cawnpore*: Small is referring to the massacre of women and children prior to the relief of the city in mid-July 1857. The European inhabitants, after defending themselves against the besieging Sepoys, were said to have capitulated on the sworn promise of Nana Sahib that he would allow them to withdraw unmolested. As they were embarking, they were set upon and indiscriminately slaughtered. The women and children were carried back to Cawnpore. Lord Roberts recorded in his *Forty-one Years in India* (1911): 'It is impossible to describe the feelings with which we looked on the Sati-Choura Ghat, where was perpetrated the basest of all the Nana's base acts of perfidy; or the intense sadness and indignation which overpowered us as we followed the road along which 121 women and children (many of them

well born and delicately nurtured) wended their weary way. After their husbands and protectors had been slain, the wretched company of widows and orphans were first taken to the Savada house, and then to the little Native hut, where they were doomed to live through two more weeks of intensest misery, until at length the end came, and the last scene in that long drama of foulest treachery and unequalled brutality was enacted. Our unfortunate countrywomen, with their little children, were murdered as the sound of Havelock's avenging guns was heard.'

102 *Feringhee*: an Indian expression meaning European.

103 *rajah*: a Hindu prince ruling a territory.

the Company's Raj: the dominance of the British East India Company which was established at the expense of the French East India Company by Robert Clive's victories in the Seven Years War (1756–63). For the next decade the Company controlled the government, its powers then being restricted by a series of Government of India Acts. Supreme political power was vested in a board of control, responsible to the British Parliament, while the Company retained administrative and commercial powers. In the nineteenth century these were gradually limited and the Company ceased to exist in 1873.

104 *ne'er-do-weel*: Anglicized to 'ne'er-do-well' in some English editions, but the jocular Scots is clearly ACD's, if a trifle cosmopolitan for Jonathan Small (but there were many Scots in the British Army from whom to adopt the usage).

moidores: formerly a Portuguese gold coin which came into use in England in the eighteenth century. The word survived as a name for the sum of 27 shillings which was the coin's approximate value.

105 *Rajpootana*: (more correctly, Rajputana) a region in north-western India which was settled in the seventh century by the Rajputs, later an affiliation of twenty-one Indian states.

108 *Wilson*: Brigadier General Archdale Wilson (1803–74), commander of the Bengal artillery at the outbreak of the Mutiny. He captured Delhi on 20 Sept. 1857.

Sir Colin: Sir Colin Campbell relieved Lucknow on 17 Nov. 1857.

Nana Sahib: properly, Brahmin Dundhu Panth (?1820–59?), adopted son of the ex-peshwa and head of the Maratha Confederation who was deprived of his title and pension on his adoptive father's death in 1853 and became the mutinous Sepoys' leader in Kanpur. He escaped into Nepal and died in the hills.

Colonel Greathed: not the future Major-General William Wilberforce Harris Greathed (1826–78), although he had twice carried despatches from Agra to Meerut through mutineers' lines in 1857, but his brother, Colonel (later Sir) Edward Harris Greathed (1812–81), who entered Agra on 10 Oct. 1857.

Pandies: nickname for the mutineers, from the name of Pande, who was a ringleader at the beginning of the rebellion.

109 *Blair Island*: (correctly, Port Blair) on the 49-mile long middle island of the five chief islands, South Andaman, whose Mount Harriet is 1,193 feet high.

110 *soldiers used always to lose*: General Drayson, under whose influence ACD came during his time in Southsea, was a relentless authority on why those around him lost at cards.

111 *send in my papers*: resign my commission.

112 *to the Governor-General?* : presumably, even the convicts on Andaman knew that in August 1858 the Governor-General was abolished and replaced by a Viceroy, but Charles John Canning, Earl Canning (1812–62) was retained and Small was simply giving him his old title.

114 *Rutland Island*: the southernmost and smallest (11-mile long) of the five chief Andaman Islands, collectively known as the 'great Andaman'.

chokey: slang for jail or prison, from the Hindi *chauki*.

115 *Pathan*: a member of one of the native tribes of Afghanistan.

119 *Schade dass die Natur nur einen Mensch aus dir schuf, Denn zum würdigen Mann war und zum Schelmen der Stoff*: 'Nature, alas, made only one being out of you although there was material for a good man and a rogue.' The lines are taken from *Xenian* (1796), a collection by Goethe and Schiller.